CLIMATE INSURGENCY

CLIMATE INSURGENCY
A Strategy for Survival

JEREMY BRECHER

Paradigm Publishers
Boulder • London

Copyright © 2015 by Paradigm Publishers

Published in the United States by Paradigm Publishers, 5589 Arapahoe Avenue, Boulder, CO 80303 USA.

Paradigm Publishers is the trade name of Birkenkamp & Company, LLC,
Dean Birkenkamp, President and Publisher.

Library of Congress Cataloging-in-Publication Data
Brecher, Jeremy.
  Climate insurgency : a strategy for survival / Jeremy Brecher.
     pages cm
  Includes bibliographical references and index.
  ISBN 978-1-61205-820-7 (hardcover : alk. paper) —
ISBN 978-1-61205-823-8 (library ebook)
  1. Climatic changes—Prevention. 2. Global warming—Prevention.
3. Environmental degradation—Prevention. I. Title.
  QC903.B74 2015
  363.738'747—dc23
                                                          2014029662

Printed and bound in the United States of America on acid-free paper that meets the standards of the American National Standard for Permanence of Paper for Printed Library Materials.

19 18 17 16 15   1 2 3 4 5

Contents

Acknowledgments

This book grows out of my work with the Labor Network for Sustainability. I thank my LNS colleagues Becky Glass, Brendan Smith, Joe Uehlein, and the late Tim Costello for their invaluable support, for their contribution to my education, and for the opportunity LNS has given me to devote my work life to climate protection.

I thank all those who have read and commented on drafts, especially John Feffer and Peter Certo of Foreign Policy in Focus and Craig Collins of California State University, Ellen Cantarow, Steve Fraser, Michael Pertschuk, Charles Lindblom, May Boeve, and Richard Falk, who persuaded me that I had to write this book. Mary Christina Wood has generously and patiently guided me in matters pertaining to the public trust. Shannon Gibson provided me valuable materials on the climate justice movement.

Jennifer Knerr, my editor at Paradigm Publishers, has not only provided valuable recommendations for the revision of this book, but has also provided ongoing encouragement for my work that has helped give me the confidence to continue with it.

Jill Cutler has not only read and edited innumerable drafts of this book, but has also shared the joys and tribulations of the life from which it springs.

INTRODUCTION

The 2013 Fifth Assessment Report of the Intergovernmental Panel on Climate Change [IPCC] confirmed that humans are destroying the earth's climate, but it also revealed something that should be even more alarming: Twenty-five years of human effort to protect the climate have failed even to slow that destruction down. On the contrary, annual carbon emissions from burning fossil fuels have risen 60 percent since the release of the first IPCC report in 1988.[1] In April 2014, carbon dioxide levels in the atmosphere reached their highest level in 800,000 years.[2]

Climate Insurgency: A Strategy for Survival lays out an approach to protecting the earth's climate that it describes as a "global nonviolent constitutional insurgency." It starts with a brief history of official climate protection efforts "from above" and nongovernmental ones "from below" that explains why climate protection efforts have so far failed. Then it proposes a global insurgency to overcome that failure, offers a potential legal basis for a climate-protecting insurgency, indicates how to make national economies climate-safe and justly distribute the global costs and benefits of climate protection, and lays out a strategy for imposing climate-safe practices on governments and economies.

When scientists first established that human burning of fossil fuels was causing global warming, the solution seemed obvious and at hand: National governments needed to agree to make modest annual reductions in the total amount of greenhouse gases (GHGs) that were emitted into the atmosphere. Global negotiations had recently resulted in an agreement to phase out another pollutant, chlorinated hydrocarbons, that had been causing a hole in the ozone layer. Greenhouse gases seemed controllable through similar measures taken by the same kind of institutions.

Twenty-five years later, however, there is no binding agreement to limit GHGs; emissions have reached a level that guarantees a warming of at least 2°C (3.6°F). Despite extreme storms, floods, droughts, arctic melting, desertification, fires, and other indicators of apocalyptic climate change, emissions of carbon and other GHGs continue to rise and are projected to go on rising, which will lead to still more devastating climate change. Yet there is no significant limitation on further emissions, inadequate analysis of the reasons for this failure, and little plausible strategy to overcome it.

Advocates of climate protection have been repeatedly defeated or forced to accept inadequate measures by visible opponents. These are generally collections of fossil fuel producers and users, politicians influenced by them, and right-wing ideologues who oppose climate protection on the grounds that it requires public interference with the private economy.

But why can a straightforward solution to a problem—one that promises such devastation for every person on earth—be blocked by such a relatively narrow collection of forces? Are there deeper structural factors that make climate protection so difficult? And if so, how can they be overcome? These are the questions explored in *Climate Insurgency*.

The failures of the past quarter century are not what most climate protection advocates expected. From the scientific confirmation of global warming in the 1980s, they had laboriously built institutions like the United Nations Framework Convention on Climate Change [UNFCCC] and the Intergovernmental Panel on Climate Change [IPCC] and had painstakingly constructed a consensus among scientists, government leaders, and UN officials around the policies defined as necessary by the IPCC. The UN "framework agreement" was followed by the Kyoto Protocol and the Bali Road Map for the Copenhagen climate summit. Based on the compelling arguments of the British government's Stern report on the economics of climate change, many global business leaders signed on to climate protection policies. Many national governments initiated policies and passed legislation to reduce GHGs. The world seemed to be proceeding on a rational, if tardy, course to address climate change.

With the collapse of the Copenhagen climate summit in 2009, it became evident that the entire process had been little more than a charade in which world leaders, governments, and business pretended to address climate change while pursuing policies that pour ever more GHGs into the atmosphere. Copenhagen revealed a collection of greedy, advantage-seeking institutions whose leaders were unable to cooperate even for their own survival.

It is doubtful that in the course of twenty-five years the official climate protection process has led to any significant reductions in greenhouse gas emissions. A UN analysis showed that, in the unlikely event that all nonbinding national pledges to cut emissions were fulfilled, the result would still be a devastating 3°C warming of the earth.[3] In 2013, carbon in the atmosphere reached 400 parts per million (ppm)—already far above the 350-ppm level that climate scientists regard as the safe upper limit.

In response to the failures of the official climate protection process, an independent climate protection movement has emerged. It is not controlled by any national or special interest. Instead, it has organized globally and has demonstrated the capacity to act globally, exemplified by the first International Day of Climate Action in 2009 that CNN called "the most widespread day of political action in our planet's history." This movement has made a global icon of what needs to be done—reduce carbon in the atmosphere to less than 350 ppm. It has broken out of the constraints of lobbying and demonstrating within a legal framework set by governments by instead adopting civil disobedience as an important and legitimate part of its strategy. It has challenged the governments that permit climate destruction, the fossil fuel–producing and –using industries that conduct it, and the corporations and other institutions around the world that collude with it.

In spite of these advances, the movement's ability to sharply reduce GHG emissions and establish climate-safe levels of carbon in the atmosphere has so far proven minuscule. Accordingly, there is a search underway to develop more effective strategies for climate protection. Some have advocated some kind of national or global revolution to overthrow the powers that perpetuate climate destruction. Others have called for building resilient local communities that can withstand climate change. Some have advocated an ecological socialism as the solution, with others proposing a purer market that charges polluters for the social cost of their emissions.

While innumerable books present the science of climate change, argue about its validity, and issue policy prescriptions for dealing with it, *Climate Insurgency* addresses the question of why climate protection has failed and proposes a possible strategy to overcome its failure. I do not claim that this is the only or even the best such strategy. Rather, I offer it in the spirit of a "thought experiment." I will be delighted if its proposals are rejected, as long as they help stimulate a successful search for better ones.

Twenty-five years ago, during the "hot summer" of 1988, my milk deliveryman said to me, "I talk with old-timers who can't remember anything like it in sixty, seventy years. It's probably this 'greenhouse effect.' If you ask me, it's a warning. All the poisons we're putting into the air and the water—if we don't get our act together, we're going to make the earth a place people can't live on." I sat down and wrote a commentary in the *Chicago Tribune* prophesying a "second ecological revolution."

> The "greenhouse effect" resulting from burning too much fuel worldwide causes droughts in many parts of the world and the heat wave my milkman so clearly perceives as a

warning. That warning is already evident to many and is rapidly becoming evident to many more: Damage to the global environment threatens the basic conditions on which life depends and poses a clear and present danger that requires a global response.

The second ecological revolution, I noted, will have to impose its agenda on governments and businesses. It will have to say that preserving the conditions for human life is simply more important than increasing national power or private wealth.[4]

Unfortunately, my prophesy was wrong, or at least a quarter century premature.

Since that time, I have closely followed and sometimes participated in efforts to protect the climate—and their failure. I helped found the Labor Network for Sustainability and was arrested at the White House in the early protests against the Keystone XL pipeline.

I have also long studied the process by which people who appear powerless and divided sometimes come together to act collectively on problems they share—a process I call "common preservation." In a book called *Save the Humans? Common Preservation in Action*[5] and a forthcoming sequel I explore how we might apply that process to today's global economic and environmental crises. *Climate Insurgency* presents climate change as a problem that requires new forms of common preservation.

Part I, "Why Climate Protection Has Failed," offers what may be the first historical account of both governmental and nongovernmental climate action, presenting their accomplishments and the reasons they have not been more successful.

Chapter 1, "Discovering the Inconvenient Truth," describes how the new problem of climate change and what to do about it entered our individual and collective awareness.

Chapter 2, "Climate Protection from Above," gives a capsule history of governmental and intergovernmental climate

protection efforts, from the confirmation that GHGs are causing global warming, up to the vacuum in climate negotiations
and legislation since the collapse of the Copenhagen climate
summit in 2009.

Chapter 3, "Climate Protection from Below," describes the
evolving climate protection efforts of citizens, nongovernmental
organizations, and social movements.

Chapter 4, "What Climate Protectors Have Accomplished,"
sums up the achievements so far of climate protection from above
and from below and indicates what these efforts can contribute
to future climate protection strategy.

Chapter 5, "Why Climate Protection Has Failed," examines
some of the barriers that have prevented us from meeting the
obvious necessity of radically reducing GHG emissions. Some
of these roadblocks are widely recognized, such as the power
of the fossil fuel industry. Some are more difficult to perceive,
however, because they are deeply embedded in a world order
we often take for granted. Identifying these deeper forces is
crucial for understanding and overcoming our current climate
protection impasse.

Part II, "A Plausible Strategy for Climate Protection?" lays
out a possible future course for the climate protection movement
that grows out of its history but embodies a more confrontational
relation to the existing world order, a new legal grounding, and
novel approaches to the transition to a climate-safe economy,
the allocation of global responsibilities for climate protection,
and political action.

Chapter 6, "A Global Nonviolent Constitutional Insurgency," indicates an alternative to either functioning as a pressure group within existing political frameworks or some kind of
violent revolution. It proposes that the global climate protection
movement become a global insurgency using nonviolent direct
action to challenge the legitimacy of existing authorities. It
argues that such an insurgency can be justified not only morally,

but also practically as a means to contest the authority of governments that are failing to perform their most fundamental duties to those they claim to represent.

Chapter 7, "Climate Protection as a Legal Duty," explains the basis for asserting that governments have a legal obligation to prevent destruction of the earth's climate, based on the legal principle known as the public trust doctrine. This principle states that the atmosphere is part of the common property of the people, for which governments are not owners but trustees. Lawsuits are currently being brought asking courts to define the public trust obligations of state and national governments to protect the atmosphere and to require them to meet their public trust duties. Whether or not courts agree, the public trust doctrine can provide a basis for a popular insurgency to demand that governments meet those responsibilities.

Chapter 8, "Making a Country Climate-Safe," looks at how countries could actually meet their responsibilities to reduce GHGs to safe levels. While various policy approaches have been endlessly debated, sufficient GHG reduction is likely to require a combination of different approaches, including direct government planning, investment, regulation, and enforcement; market mechanisms such as carbon taxes and cap-and-trade systems; and massive participation in local grassroots initiatives. The transition to more effective approaches will require powerful government agencies to direct and enforce change, in some ways similar to those that led the conversion to war production during World War II. This, in turn, will require instruments of accountability to make sure such powers are used to achieve their intended purpose without being perverted to other ends. It will also require policies to ensure that climate protection will mean a better life for ordinary people, thereby assuring continuing popular support.

Chapter 9, "A Global Trust Fund for the Global Public Trust," examines changes that will be necessary at the global level to make the transition to climate safety. Such changes include the means to gather and allocate resources for climate protection globally; vehicles for ensuring effective national climate protection policies; and ways to make climate protection acceptable by using it to support a prosperous, job-creating global economy.

Chapter 10, "Movement Enforcement of Public Trust Duties," looks at how a climate-protecting insurgency can implement its goals. It discusses how the movement can redefine climate destruction as committing waste against the public trust, develop the independent power to protect the global atmospheric commons, work cooperatively with noninsurgent allies, and reverse the climate-destroying dynamics of the world order.

Chapter 11, "Overcoming the Obstacles to Climate Protection," reviews how the strategy presented here can help overcome the forces described in Part I that have so far perpetuated destruction of our climate.

The Conclusion aims to describe the strategy as a coherent whole.

Although this book attempts to address climate protection in a global perspective, it inevitably reflects and is limited by my personal experience as an American.

I hope readers will examine the strategy proposed here critically. But I also hope they will either correct its flaws or develop a better alternative. Climate protection can't wait for a perfect strategy; all of us have a duty to find the best strategy we can—and then act on it.

Part I

Why Climate Protection Has Failed

CHAPTER 1
DISCOVERING THE INCONVENIENT TRUTH

For millennia, some have said the world would end in fire; some have said in ice. The possibility of catastrophic global warming generated by human beings was clear as early as 1964, when social ecologist Murray Bookchin wrote,

> The burning of fossil fuels (coal and oil) add 600 million tons of carbon dioxide to the air annually, about .03 percent of the total atmospheric mass.... Since the Industrial Revolution, the overall atmospheric mass of carbon dioxide has increased by 25 percent over earlier, more stable levels.... This growing blanket of carbon dioxide, by intercepting heat radiated from the earth, will lead to more destructive storm patterns and eventually to melting of the polar ice caps, rising sea levels, and the inundation of vast land areas.[1]

Various individuals, groups, and institutions became aware of global warming on varying timetables.[2] For the first two decades following Bookchin's prediction, global warming remained an arcane discussion among scientific specialists. Conflicting evidence indicated that greenhouse gases would warm the world or that, conversely, air pollution would block sunlight and therefore cool the earth down. Starting in the 1980s, converging lines of evidence established with growing certainty that global warming would be our fate. The international community of climate scientists concluded that it was their responsibility to inform governments of the looming threat. In 1988, the climate scientists persuaded the World Meteorological Organization and the United Nations Environment Program to establish the Intergovernmental Panel on Climate Change (IPCC). Some scientists also began trying to inform the public about global warming, and reports about it began appearing in the media; but "most scientists felt more comfortable sending rational appeals through channels to government officials."[3]

In the United States, the August 1988 heat wave projected global warming into public discussion. The number of articles on global warming in US newspapers increased tenfold between 1987 and 1988. Polls conducted during 1988 found that most Americans said the greenhouse effect was very serious or extremely serious and that they worried a fair amount or a great deal about global warming. Less than a fifth said that they worried "not at all" or had no opinion.[4]

However, people had to connect many dots in order to identify climate change as a problem of common human preservation. There were the facts, which on their face appeared ambiguous: When a heat wave struck, discussion of global warming soared. When a snowstorm hit, climate skeptics proliferated. Then there was the scientific theory, which was often opaque to anyone except specialists; moreover, this concept depended on meteorological complexity theory that was often counterintuitive to the public, generating paradoxical predictions like intensified snowstorms in some locations due to global warming.

There was also the question of what climate change would mean for individuals and groups. How soon would its impact be felt, would it just be a trivial change, would it devastate some areas and barely affect others, or would it mean a universal human catastrophe?

There were many ways to avoid thinking about global warming. Perhaps most common was simply to disregard it—something not worth thinking about amid the welter of interests and concerns that people faced daily. Alternatively, it could be despairingly accepted—an inescapable inevitability like death and taxes. It could also be flatly denied. Climate skepticism and denial were promoted by deliberate campaigns funded by fossil fuel corporations and investors, who consciously imitated the success of the tobacco industry in blocking tobacco regulation for decades by raising doubts about the science linking smoking and health.[5] In the United States the oil companies took advantage of the populist antiscientific disparagement of know-it-all elites, which echoed the strand of anti-Darwinism in American culture: Common sense showed that human beings couldn't possibly have an effect on the entire global atmosphere, any more than humans could have evolved from apes.[6] Fears of anthropogenic climate catastrophe could be viewed as an absurd violation of the well-established presumptions that technology, market economies, and economic growth lead to beneficial progress.[7] Conservative ideologues not only denied the reality of climate change, but attributed the push to recognize and act on it as a plot to impose internationalist and socialistic policies on the world.

Even for those who accepted the reality of climate change, there were many ways of interpreting it that simply did not lead to efforts for common preservation. Policymakers often regarded climate change and climate change policy primarily in terms of its effects on economic development and international economic competition. Fossil fuel corporations largely regarded consciousness of global warming as a threat to their profits. Some unions regarded climate protection efforts as a threat to jobs. Consumers could see climate protection as something that would add to the cost of the goods they bought. The media, the general public, and environmentalists themselves often defined global warming as an "environmental issue," likely to have more impact on polar bears than on people.

Something so apocalyptic in scope also led to theological interpretations.[8] Some thought that global warming was God's punishment for hubris or sin or that it was God's will, which it was our duty not to understand or question but to accept. Alternatively, others thought that world-destroying climate change was inconceivable, because a benevolent God would never allow it. Restrictions on human exploitation of nature were sometimes portrayed as almost blasphemous, since God intended the earth for the use of man. In some views, stewardship of the natural world could be seen as man's duty.

As the reality of climate change was increasingly recognized, another process was necessary to transform it from a brute fact to a motivation for action: It needed to pose a problem. Climate change violated expectations of the continuity of life. It threatened almost everything that people valued—personal well-being, families, communities. It posed a threat to the well-being of future generations and of people's own children. It also raised profound ethical issues, as climate change and efforts to adapt to it impacted different social groups and generations in very different ways. The specter of climate change also violated widely accepted norms, generating outrage as individuals and corporations profited from destroying the environment and those supposedly responsible for the security and well-being of their populations acted in utter disregard of the threat. It represented something that nobody wanted, but that had been imposed upon them nevertheless.

These emerging recognitions provided powerful motivations to define climate change as a problem requiring a solution. It was an obstacle to be overcome; fixing it was a need to be fulfilled and halting it was in people's most profound interest. A 1999 study based on poll data and focus groups noted that in one survey "three out of four Americans" believed that "the earth's atmosphere is gradually warming as a result of air pollution and that, in the long run, this could have catastrophic consequences." Most, however, felt they were personally powerless to halt global warming, and that indeed the problem was insoluble. "People literally don't like to think or talk about the subject." Their concern "translates into frustration rather than support for action."[9]

The question of what needed to be done appeared relatively simple: reduce the greenhouse gases humans put into the atmosphere. The question of how to do that turned out to involve very extensive economic, social, and political changes, however. Those changes pose many stumbling blocks unless they are considered not just as choices of individuals or policies of governments and corporations but instead as a collective human problem of common preservation.

Climate protection has sometimes been portrayed as something that individuals could fix by individual action. Al Gore's documentary *An Inconvenient Truth* concluded with a list of actions individuals could take to reduce their GHG emissions, but it was an illusion to believe that climate change could be effectively addressed by individuals acting singly. For example, even if everyone on earth turned off the electric light when they left a room, it would have a negligible effect on climate—the burning of fossil fuels in power plants or the emissions resulting from an automobile-based transportation system would far outweigh the effect of such individual lifestyle changes.

Nor could climate change be fixed by extending the prevailing strategies of dominant institutions. Some economists, it is true, argued that economic growth would create a rising standard of living that would allow citizens to become more concerned with such seemingly intangible matters as the environment, and would create a surplus that would allow expenditure on pollution control. In fact, prevailing forms of economic growth greatly exacerbated GHG emissions. Moreover, while militaries increasingly recognized climate change as a national security threat, it was not a threat that could be dealt with by sending an army to invade a GHG-emitting country.

As localities around the world were devastated by storms, floods, droughts, and other effects of climate change, clamor arose for dams, dikes, reservoirs, and other ways of protecting against climate change effects—what is known as adaptation. After Superstorm Sandy in 2012, for example, New York City initiated a multibillion-dollar infrastructure project to protect against future storms.

These adaptive measures had contradictory effects. On the one hand, many people concluded that if public officials were actually willing to spend significant money on climate change adaptation, it probably meant that the danger was real. On the other hand, it led to a belief that adaptation makes measures to protect the climate unnecessary. Not surprisingly, fossil fuel companies became enthusiastic advocates of this latter view. Exxon Mobil CEO Rex Tillerson, for example, argued that efforts to address climate change should focus on engineering methods to adapt to shifting weather patterns and rising sea levels rather than trying to eliminate the use of fossil fuels. "Changes to weather patterns that move crop production areas around—we'll adapt to that. It's an engineering problem and it has engineering solutions."[10]

While adaptation efforts could no doubt protect against some of the local and temporary effects of climate change, most would be ineffective or counterproductive beyond a certain point. Dams and seawalls might reduce mild flooding, but they were unlikely to do much good in the face of a major rise in sea level. Individuals might adapt by using more air conditioning, but the result would be more carbon pollution overall. Adaptation measures also raise significant social justice issues, as they are generally far more accessible to affluent communities with substantial resources than to impoverished areas. Climate protection advocates have generally supported adaptation measures primarily when they also help reduce GHG emissions.

Indeed, by far the most important and effective form of adaptation to climate change is to reduce dependence on fossil fuels. Dependence on increasingly scarce and therefore increasingly costly coal, oil, and gas is already taking a heavy toll on producers, consumers, and communities, and could well lead to gradual or rapid economic decline. That dependence is also leading to ever more dangerous and destructive forms of energy extraction, such as tar or bituminous sands oil, natural gas fracking, and mountaintop-removal coal mining. Our ever more costly and destructive fossil fuel energy system is causing not only climate catastrophe but also burgeoning economic, social, and environmental stress. Conversely, conversion to a climate-safe energy system will also provide the basis for far more secure, sustainable, and resilient communities.[11]

There are parallels between the process that created public understanding of climate change and that which led to the "ban the bomb" movement in the mid-twentieth century. The initial warnings about the atomic threat came from the atomic scientists and were not something people could derive from their personal experience. The scientists went first to government officials with their warnings. Government officials, however, regarded the arms of other countries as the significant threat, and failed to grasp the vicious circle of the arms race in which they themselves played significant roles. Attitudes changed, however, when ordinary people were able to grasp the atomic threat in terms of the impact of nuclear fallout that could affect their health and that of their children. The result was a global movement and a powerful demand that the nuclear powers halt nuclear testing and move toward nuclear disarmament.

The identification of climate change as a problem requiring a shift to a strategy of common preservation is today only in its early stages. The necessity for climate protection, combined with the inability of individual and prevailing institutional strategies to provide it, has led to an experimental testing process—a process whose story from 1988 to today we will tell in the next two chapters.

CHAPTER 2
CLIMATE PROTECTION FROM ABOVE

In 1988, as climate scientists became more certain about carbon-induced global warming and the global public grew more alarmed by extreme weather, the United Nations General Assembly designated the UN Environmental Program (UNEP) as the UN's venue for climate issues; UNEP, working with the UN's World Meteorological Organization, established the Intergovernmental Panel on Climate Change (IPCC), whose First Assessment Report rapidly established that global warming was real and probably caused at least in part by human release of greenhouse gases. In 1990, the General Assembly initiated negotiations for an international agreement to protect the climate.

In 1988, even before preliminary climate protection negotiations began, the US oil and auto industries and the US National Association of Manufacturers established the Global Climate Coalition to oppose any mandatory actions to address global warming. Housed at the National Association of Manufacturers, the coalition would spend tens of millions of dollars on advertising against international climate agreements and national climate legislation.[1] Drawing on the experience of the tobacco industry, its advertising and lobbying campaign successfully cast enough doubt on the reality of global warming to undermine US support for climate protection efforts.

An effective climate protection agreement required participation of almost all current and potential polluters—otherwise, GHG production would simply move to locations not covered by the agreement. Even in the preliminary negotiations for an agreement, the sharp conflicts that would plague the process for the next twenty years were evident, however. In a 1991 article titled "Climate Change Negotiations Polarize," Craig Collins, then a doctoral student studying atmospheric negotiations, found that after just three sessions the parties were at loggerheads. An "activist" coalition composed mostly of European nations and environmental nongovernmental organizations (NGOs) advocated an agreement that would require every nation to implement targets and timetables for GHG reductions. A "go slow" coalition led by the United States, the world's biggest emitter of carbon dioxide, opposed any mandatory cuts. Meanwhile, developing countries maintained that the GHG problem had been created by the developed countries; that limiting developing countries' carbon emissions would impede their development; and that if developing countries participated in any scheme limiting GHGs, the developed countries should pay for them to adopt a low-GHG pathway to development.[2]

Rio

The 1992 Earth Summit in Rio de Janeiro was set as the venue for signing a climate treaty. Climate protection activists hoped for an agreement that would actually start preserving the climate. However, as Collins later wrote, while "most developed nations saw climate change as a clear and present danger," the United States "refused to agree that the science justified specific, mutually agreed-upon targets and timetables for actual GHG reductions."[3] At Rio, virtually all the world's nations signed the United Nations Framework Convention on Climate Change (UNFCCC), which established a framework for future negotiations—but did nothing to protect the climate.

While the coalition supporting an effective treaty criticized the US role, it did not mobilize effective pressure to change it. According to physicist and historian Spencer Weart, industry-induced doubt about global warming and consequent lack of US support for climate protection efforts "gave other governments an excuse to continue business as usual." Politicians could claim "they advocated tough measures, casting blame on the United States for any failure to get started."[4] Collins concurred: US skepticism had "a corrosive impact on the activist policies they were prepared to adopt." Following the summit, "the EU downgraded its carbon tax proposal in part because of U.S. inaction."[5]

Nonetheless, efforts to reduce GHGs proliferated around the world. Educational campaigns urged consumers to "go green." Cities, states, provinces, regions, and corporations initiated programs of GHG reduction. Most nations developed climate protection policies of one kind or another. The UNFCCC held annual Council of Parties (COP) gatherings in pursuit of a climate treaty, scheduled for negotiation at Kyoto in 1997.

Kyoto

As the question of how to actually limit greenhouse gas emissions was posed, the division between developed and developing nations moved front and center. Developing nations argued that the developed nations were responsible for almost all the greenhouse gas emissions in history, and should bear the responsibility for correcting them. Some portrayed limits on GHG emissions as a plot to prevent poorer nations from developing by restricting their use of fossil fuels.[6] Industrial countries, especially the United States, argued that no agreement limiting GHGs would be effective if it did not also apply to the growing emissions from developing countries.

After contentious negotiations, a compromise was cobbled together and the Kyoto Protocol was signed. Its ostensible goal was to prevent the atmosphere's temperature from rising more than 2°C (3.6°F). All parties eventually accepted the formula of "common but differentiated responsibilities" for developed and developing countries, but what that would mean in practice remained unresolved. After repeated deadlocks, the final agreement prescribed binding GHG reduction targets by 2012 of 5 percent below 1990 emission levels for participating developed nations, but allowed them to exceed their targets by trading quotas or investing in GHG reduction in developing countries—what became known as "cap and trade."[7]

In 2002, British prime minister Tony Blair evaluated the effectiveness of the Kyoto Protocol: "We should recognize one stark fact: even if we could deliver on Kyoto, it will at best mean a reduction of 1 per cent of global warming." But, he warned, "we need a 60 per cent reduction worldwide."[8] Although agreed to in 1997, the Kyoto Protocol was not scheduled to go into effect until 2005. Ultimately, 192 countries ratified the Protocol, including all major countries except the United States.

The Politics of Climate

The fossil fuel industry argued that the Kyoto Protocol standards for GHG emission cuts in developed countries would raise energy costs in the United States, leading jobs to flow instead to those developing countries not required to make such cuts.[9] Much of organized labor in the United States agreed with this argument and opposed ratification of the Protocol. Even before the Protocol was signed, the US Senate passed the Byrd-Hagel Resolution, which unanimously disapproved of any international agreement that did not require developing countries to make emissions reductions or that would seriously harm the US economy. The Clinton administration signed the Kyoto Protocol knowing that it had no chance of ratification in Congress.

Meanwhile, the impacts of global warming became increasingly visible. The 1998 floods in Bangladesh inundated over 75 percent of the country and killed thousands. A 2003 European heat wave caused 15,000 deaths in France, 7,000 deaths in Germany, and killed many more throughout the rest of Europe.[10] In 2005 Hurricane Katrina devastated New Orleans, capping the worst Atlantic hurricane season on record. Global warming finally returned to America's radar screen in 2006 with Al Gore's documentary *An Inconvenient Truth*.

Much of global business also appeared to be shifting away from blanket opposition to climate protection measures. The Stern review on the economics of climate change, issued by the British government in 2006, called global warming the worst market failure in history and found that climate change would have adverse economic effects equal to those of the Great Depression and World War II combined. Many large corporations withdrew from the Global Climate Coalition and joined the newly formed Business Environmental Leadership Council, which endorsed climate science and supported binding international agreements for climate protection, albeit ones with a business-friendly approach.[11]

In most European countries, public concern about global warming was substantial and was augmented by political pressure from Green parties; even conservative leaders like the UK's Margaret Thatcher and Germany's Angela Merkel supported some kind of GHG reduction program. In 2005 the EU established a system that required permits for carbon emissions and created a market for trading the permits. Rather than selling pollution permits, however, governments gave away so many that they added up to "more than the pollution companies were actually spewing forth." According to *The Economist*, "What should have been an exercise in setting rules for a new market became a matter of horse-trading about pollution limits, with powerful companies lobbying for the largest possible allowances." The price of a permit soared to thirty euros per ton of carbon, then crashed to almost nothing.[12] Skepticism about such "cap-and-trade" systems soared.

In contrast to the rest of the world, in the United States the climate issue became politically polarized. Polls typically found Democrats twice as likely as Republicans to be worried about global warming.[13] In 2001 incoming president George W. Bush said he would never agree to the Kyoto limits on carbon emissions.[14] By 2006, polls indicated worldwide hostility to the United States as a result of its climate policies.[15]

In the United States, fossil fuel corporations and right-wing political groups promoted skepticism about the science of climate change and vociferously opposed government programs to address it. For example, the former president of the National Academy of Sciences, Dr. Fredrick Seitz—a physicist but not a climate scientist—led a campaign to discredit the "theory" of global warming. Investigative reporting revealed that Seitz had been a consultant to tobacco giant R. J. Reynolds, where he oversaw the distribution of $45 million for "research," and that in 1972 he had become a paid director and shareholder of the Ogden Corporation, which operated coal-fired power plants.[16] Many other "global warming skeptics" turned out to be receiving funding from fossil fuel companies as well, but in most cases the media presented them as legitimate proponents of a scientifically tenable position. This absurdity represented success for the underlying fossil fuel industry strategy; as an internal Exxon memo explained, "Victory will be achieved when uncertainties in climate science become part of the conventional wisdom" for "average citizens" and "the media."[17]

Nonetheless, a broad US coalition supported some kind of climate legislation. Barack Obama's 2008 presidential campaign had included strong statements on the need to protect the climate, and Democrats held majorities in both houses of Congress. The bill that emerged in Congress, known as the Waxman-Markey or the American Clean Energy and Security Act, was based on a domestic cap-and-trade system; it was so weak and provided so many sweeteners for coal and other fossil fuel companies that some environmental groups like Greenpeace and Friends of the Earth refused to support it. The legislation passed the House of Representatives in 2009 but was blocked in the Senate.[18]

The *New York Times* succinctly summarized the history of climate change legislation in the United States and the reasons for its defeat:

> Efforts to enact a carbon price in Washington have failed largely because powerful fossil fuel groups financed campaigns against lawmakers who supported a carbon tax.
>
> In 1994, dozens of Democratic lawmakers lost their jobs after Al Gore, who was vice president at the time, urged them to vote for a climate change bill that would have effectively taxed carbon pollution. In 2009, President Obama urged House Democrats to vote for a cap-and-trade bill that would have required companies whose carbon dioxide emissions exceeded set levels to buy emissions rights from those who emitted less. The next year, Tea Party groups spent millions to successfully unseat members who voted for the bill.[19]

President Obama, while calling for climate legislation, largely stood aloof as it crashed in Congress. In the 2012 presidential campaign debates, the issue of climate change was not even mentioned.

Copenhagen

The Kyoto Protocol was scheduled to expire in 2012. The Bali 2007 COP established the Bali Roadmap to negotiate a binding agreement for far greater emission cuts at the 2009 climate summit at Copenhagen—not as fast or deep as everyone might have hoped, but perhaps sufficient to prevent the worst effects of global warming. When US representatives obstructed the compromise at the last minute, they were booed and hissed. In a dramatic moment, the delegate of sea-rise-threatened Papua New Guinea said, "If for some reason you are not willing to lead, leave it to the rest of us. Please—get out of the way." The United States backed down and withdrew its opposition, marking what Weart called "a striking demonstration of the power of public opinion."[20]

Copenhagen was anticipated to be one of the most important global gatherings in world history. British prime minister Gordon Brown said, "In every era there are only one or two moments when nations come together and reach agreements that make history, because they change the course of history. Copenhagen must be such a time."[21] But in the run-up to Copenhagen, President Obama signaled that no binding agreement would be reached: On November 14, 2009, the *New York Times* announced that "President Obama and other world leaders have decided to put off the difficult task of reaching a climate change agreement," agreeing instead to "make it the mission of the Copenhagen conference to reach a less specific 'politically binding' agreement that would punt the most difficult issues into the future."[22]

The conference itself polarized into those seeking a binding agreement for deep emission cuts and those seeking to block such an agreement. An alignment of the G-77 developing countries, NGOs, and demonstrators pushed for agreement, but the major powers refused to compromise on their national economic ambitions. Ultimately leaders of the world's two largest carbon polluters, the United States and China, negotiated a voluntary, nonbinding agreement and agreed to hold their own negotiations outside the UN process with any others who chose to join them. Neither those negotiations nor the continuing annual COP sessions nor the September 2014 UN climate summit have produced any reductions in GHG emissions.

Protecting the earth's climate is in the long-term interest of all humanity. Yet efforts to cut carbon and other GHGs to a climate-safe level have been defeated for a quarter century in arenas ranging from the United Nations to the US Congress. The result has been the emergence of a very different kind of climate protection movement.

CHAPTER 3
CLIMATE PROTECTION FROM BELOW

A s the evidence of climate danger grew, more and more people began to think that climate protection was too important to leave to either the scientists or the politicians. Over time, a nongovernmental climate protection movement emerged. At first it was closely intertwined with official governmental efforts, but over the years significant elements of it became more independent. Originally the effort was largely an extension of existing environmental organizations, but eventually it came to include almost every imaginable kind of constituency and concern. It grew less by design than by new constituencies becoming conscious of the realities of climate change and drawing together to make new collective responses. While it always involved a combination of formal NGOs and less formalized networks, it became increasingly decentralized and fluid. Although the movement originally concentrated on education, lobbying, legislation, and electoral action, over the span of a quarter century its focus shifted to more direct confrontation with governments and industry.

Climate protection from below grew out of constituencies and cultures with divergent or even contradictory strategies, modes of action and organization, ideologies, and interests. It would be an unlikely candidate for a common movement if its recruits did not also face a common danger. As climate protection from above ever more clearly failed to avert that danger, climate movements developed not only in response to climate change itself, but to the official climate process and to each other.

Civil Society and Climate Action

As climate scientists became more confident that GHGs were causing global warming, environmental organizations identified climate as a "green" issue and began adding climate protection to their agenda. Environmental groups including the World Wildlife Fund, the Union of Concerned Scientists, and Greenpeace developed their own scientific expertise and began campaigning for national and international limits on GHGs. In 1987, environmental activists working with scientists initiated an appeal for US action on climate change that was signed by half the Nobel Prize winners in the United States and half of the members of the National Academy of Sciences.[1]

That same year, environmental organizations worldwide created the Climate Action Network (CAN), which connected NGOs in North America, Asia, Africa, Latin America, and eastern and western Europe. CAN aimed to persuade governments to limit GHGs to a safe level, emphasizing the primary responsibility of industrialized nations to take the lead. CAN served both as a vehicle for global coordination and information exchange among NGOs and as an advocate in its own right, demanding that countries live up to their responsibilities and focusing global public opinion on the countries' failure to do so. During international climate negotiations, CAN published an influential daily newsletter, *ECO,* which ricocheted rapidly around the world, defining the issues of the day and exposing attempts by national governments to evade them.[2]

Both CAN and its affiliates mobilized support for the United Nations Framework Convention on Climate Change (UNFCCC) process and for a climate treaty with binding targets and timetables. CAN became the official vehicle for representing NGOs within the UNFCCC process. The number of NGOs officially admitted rose from 174 at UNFCCC Council of Parties (COP) 1 to 1,406 at COP 16.[3] The formal inclusion of NGOs in such official international activities was a novelty, but one that was also occurring in other international venues, reflecting the rising importance attributed to "civil society" in global discourse. Indeed, CAN represented a drawing together of climate protection "from above" and "from below" in complex interaction without strong differentiation between the two.

Meanwhile, people around the world began taking initiatives for climate protection at a local level. Annual International Days of Climate Action featured mass marches, mass bike rides, and large concerts around the world.[4] When the United States refused to ratify the Kyoto Protocol, Seattle and hundreds of other American cities announced that they would nonetheless voluntarily meet its reduction targets. Many campaigns developed that linked local environmental pollution to GHG emissions. For example, environmental organizations and community activists in hundreds of local communities in the United States and elsewhere campaigned to prevent the building of new coal-fired power plants and to shut down existing ones, both because of their effect on local health and environment and because of their contribution to global warming.[5] These activist groups fought what came to be referred to as "extreme energy" extraction, such as natural gas fracking, mountaintop-removal coal mining, and mining of tar sands oil.

Climate action also spread through a variety of social sectors. An international Youth Climate Movement emerged, based on the self-expressed interest of young people in the future of the planet, to articulate youth interests at COP gatherings and organize "Power Shift" conferences around the world. Young people also demonstrated and brought suits against governments for permitting the climate to be destroyed in ways that laid waste to the rights of present and future generations.[6]

Many religious leaders, congregations, and denominations similarly took on responsibility for addressing climate change. In some areas, church bells were tolled 350 times to proclaim the importance of reducing carbon in the atmosphere to 350 ppm. A statement signed by the Dalai Lama, the Patriarch of Constantinople, and many other world religious leaders called for sharp reductions in GHG emissions.[7] Some churches and denominations moved to divest their holdings in fossil fuel corporations. Moreover, some Christian evangelicals abandoned their established rejection of environmentalism and instead began calling for responsible environmental stewardship.

Indigenous peoples around the world also emerged as opponents of climate change. Indigenous beliefs about the natural world often condemned destructive exploitation of the environment; indigenous lands and communities were often the targets of highly destructive fossil fuel extraction. Indigenous peoples initiated opposition to extraction of Alberta tar sands and formed the Idle No More movement that challenged energy extraction across Canada. The Indigenous Environmental Network linked environmental concerns around the world and became a significant voice in international climate negotiations.

Organized labor played an ambiguous role in the climate protection movement. The International Trade Union Confederation, which represents the 175 million members of 315 affiliated organizations in 156 countries and territories, consistently backed and lobbied for international climate agreements with binding targets and timetables. It called for a "just transition" that would protect the well-being of workers and communities affected by the necessary shift to a climate-safe economy. Although many unions around the world launched climate protection efforts in their own workplaces and communities, unions in fossil fuel–producing and –using industries often felt their jobs threatened by climate protection policies like cap-and-trade systems and carbon taxes. This outlook was particularly common in the United States and in coal-dependent countries like Poland. The AFL-CIO, a federation that includes most unions in the United States, opposed the Kyoto Protocol and never supported the targets and timetables for GHG reduction called for by the IPCC. However, some unions, especially in the transportation and service sectors, were strong advocates of climate protection.

Climate Justice

After the turn of the twenty-first century, climate activism underwent a change described by international relations scholar Shannon Gibson as going from "large, well-funded, Northern nongovernmental organizations (NGOs) focused on single issues, science-based technocratic approaches to lobbying, and media and family-friendly campaigns to 'save the [insert animal here]'" to "an ever-shifting constellation of environmental, human rights, debt-free and social justice NGOs, social movements, grassroots organizations and autonomous activists simultaneously coordinating and competing at the transnational level on various campaigns from biological diversity to climate change."[8]

As the UNFCCC negotiations moved toward implementing a climate protection system based on cap and trade and carbon offsets, the civil society organizations represented in CAN began to divide. Major groups like the Sierra Club, the World Wildlife Federation, and Environmental Defense, as well as CAN itself, supported such "market-based mechanisms" if they implemented binding targets and timetables for GHG reduction. In contrast, groups like Friends of the Earth, Focus on the Global South, Carbon Trade Watch, the Center for Environmental Concerns, the Philippines Freedom from Debt Coalition, and the World Rainforest Movement rejected the mechanisms as both ineffective and unjust. In the aftermath of the Bali COP these dissenting groups created a new network, Climate Justice Now! (CJN!).

A principal target of the nascent climate justice movement was a program called Reducing Emissions from Deforestation and Forest Degradation (REDD). While CAN supported REDD, CJN! characterized it as "land-grabbing" schemes that would hurt forest-dwelling and Indigenous communities.[9] World Rainforest Movement spokeswoman Ana Filipini charged that, "Carbon finance mechanisms in developing countries result in forests being transferred or sold to large corporations which hope to acquire profitable 'carbon credits' associated with those forests at some point in the future."[10] The Indigenous Environmental Network (IEN) published a booklet titled *REDD: Reaping Profits from Evictions, land grabs, Deforestation and Destruction of biodiversity.*[11]

The CJN! network was also a response to what it identified as bias in the UNFCCC and CAN itself. As one of its organizers told CJN! activist-scholar Scott Byrd, "Before Bali, there was very little in the way of organizations representing affected peoples from the South within the UNFCCC; Indigenous groups, farmers, forest and fisher folk were simply not part of the process. REDD brought these groups to the forefront as affected populations, but there was no space for them to coalesce around, so the network in the beginning was simply a way to create space for these voices, for opposition from the south to be heard."[12] At COP 14 the CJN! was officially recognized as an environmental NGO network by the UNFCCC, and CAN began splitting its observer badge quota, meeting space, and press conference slots with CJN.[13]

"Climate justice" soon emerged as a common frame for a range of perspectives critical of official climate protection efforts.[14] Shannon Gibson, who studied the emergence of CJN! firsthand, described the tension between "those who believe that global economic growth and development are beneficial and/or inevitable and thus should be harnessed to protect the environment" and "those who view the growth model as unsustainable and dangerous for human and other species' long-term survival."[15] Advocates of the climate justice frame focused on the relation of climate change to economic and social justice, particularly the disparities between the global North and South. It advocated "system change" as opposed to "traditional governmental ways of dealing with climate change."[16] Through these framing efforts, "climate justice" soon became "a global rallying cry, shifting activists' criticisms away from technocratic claims that targeted policies, negotiators, and specific governments toward a more antisystemic approach, which criticized developed countries writ large, the UNFCCC and various other forms of global governance, and neoliberal capitalism in the context of climate change."[17]

The climate justice frame drew on a "radical environmentalism," which held that the "enduring power structures of sovereignty, capitalism, scientism, patriarchy and even modernity generate and perpetuate the environmental crisis while consolidating structural inequalities between the global North and South."[18] Gibson discerned various "transgressive frames" within radical environmentalism, including "indigenous cosmology, deep ecology, social ecology, political ecology, environmental justice, ecofeminism, and eco-anarchism."[19]

As the Copenhagen climate summit approached in late 2009, many of the tensions within climate protection activism came to a head. CJN!, joined by activists from the "antiglobalization" or "alter-globalization" movement, made plans to confront the official meetings.[20] Many groups in CJN!, however, opposed using direct action at the COPs.[21] So social justice and anarchist groups, primarily from Europe, formed the Climate Justice Alliance (CJA) to storm the UNFCCC and establish a "people's assembly."[22] CJA was composed of "horizontally networked autonomists and alter-globalizationists" who "privileged civil disobedience and disruptive direct action over petitioning and lobbying."[23]

As the official COP 15 negotiations sputtered, civil society groups conducted the largest protests in the UNFCCC's fifteen-year history. On December 12, between 50,000 and 100,000 people marched to the Bella Center where the COP negotiations were held, demanding a "fair, ambitious and legally binding treaty to avert catastrophic climate change."[24] The legally authorized demonstration was organized by CAN, environmental NGOs, and new Internet-based climate groups like TckTckTck and 350.org. It brought together "hundreds of organizations, climate activists, human rights campaigners, and unaffiliated protesters from dozens of countries."[25]

On December 16, a very different protest, dubbed Reclaim Power, targeted the Bella Center. It was the result of a compromise between CJN! and CJA. As one activist explained, "Those who've worked on the inside of the summit for a long time agreed to a more confrontational politics, those who tend to oppose global summits in their entirety agreed on a more nuanced perspective, and not to lobby to shut down the entire summit. The compromise was to disrupt on only one day."[26]

Outside the Bella Center, 3,000 activists led by CJA tried to storm the Center. Inside, CJN! members led a noisy walkout of over 300 UNFCCC-accredited observers, "charging past government delegations and the media and overwhelming the computerized checkout system in an effort to join the throng of protestors outside trying to breach the Center's perimeter." Protesters on both sides were blocked from meeting in the middle by police forces. Many were beaten, pepper sprayed, and arrested.[27] Amidst the collapse of the climate talks, the confrontation received worldwide headlines.

In the aftermath of the Copenhagen summit, Bolivian president Evo Morales convened a World People's Conference on Climate Change and the Rights of Mother Earth (WPCCC). An estimated 35,000 people from 140 countries gathered in

Cochabamba in April 2010 to "analyze the structural and systemic causes that drive climate change and propose radical measures to ensure the wellbeing of all humanity in harmony with nature."[28] Its call to recognize the "Rights of Mother Earth" received wide attention, but there was little follow-up.[29] CJA dissolved a year later.[30] CJN! became largely inactive after 2012, although networking among its constituents has continued.[31] However, the influence of the climate justice frame has been ubiquitous. For example, the Climate Action Network reversed its position on the rights of indigenous communities "due in part to the success of CJN!."[32] Local and national climate justice organizations, such as the Climate Justice Alignment in the United States, continue to raise issues of social and economic justice in relation to climate change.

The climate justice movement was quite successful in focusing attention on the ways that programs ostensibly aimed at climate protection could in fact be used instead to increase the exploitation of poor and oppressed people. It also had several significant problems, many not of its own making. The various forces it appealed to in the developing world did not have uniform interests. The UN's Least Developed Countries, island countries, and other low-lying countries, many of them extremely poor, faced an immediate threat from the impacts of climate change and demanded strong restraints on GHG emissions. However, rapidly developing countries like the so-called BRICS (Brazil, Russia, India, China, and South Africa) were deeply concerned that they retain their right to use fossil fuels. Nor did developing nations' governments necessarily represent the interests of all their people rather than their own elites. At one CJN! event a Malay indigenous activist challenged a previous speaker who had said that "developing countries" are "our allies in the negotiations." "I was really having shivers because these are also the very countries that are violating our rights!"[33]

While the climate justice movement effectively criticized mainstream proposals for climate protection, it had difficulty formulating a program to actually restrain global warming. The network facilitators and participants "found it easier to build consensus around what they were all against rather than what they were for (potential propositions to ameliorate climate change or address the effects)."[34] Indeed, Scott Byrd found that "CJN! organizers orchestrated deliberate attempts to suppress prognostic framing during organizing meetings and discussions."[35]

Finally, there was an underlying strategic problem for climate justice efforts. If the world's dominant governments had been unequivocally committed to climate protection, a threat to discredit the UNFCCC process unless justice concerns were met could have had powerful leverage. Since the major governments were ambivalent about serious climate protection, however, attacks on the process may actually have made it easier for them to resist effective climate protection.[36]

Nonetheless, the influence of the climate justice frame can be measured by the fact that most climate activists worldwide today might well describe themselves as part of a "climate justice movement."

The Rise of Direct Action

Historically, action to protect the environment has often taken the form of disobedience to established authority. Indigenous people around the world have protected traditional ways of life by eluding or resisting colonial and other governmental authority; examples range from the Maori resistance to British encroachment in Parahaka in late-nineteenth-century New Zealand to the Ogani resistance to foreign oil companies in the Niger Delta in the 1990s and beyond. Greenpeace, whose first action in 1969 was to sail a ship into a restricted zone to protest environmentally destructive nuclear testing—and which became famous through its disruption of whale hunting—has continued its dramatic confrontations on land and sea to this day, for example, with its 2014 protest blocking an Arctic oil tanker in Holland, which led to forty-four arrests. The campaign to block clear-cutting and protect old-growth forests in the western United States involved tree sitting, blocking logging roads, and street protests, as well as various forms of so-called "monkey-wrenching" like driving nails into trees to break saw blades.

From the confirmation of GHG-caused global warming to the turn of the twenty-first century, however, civil disobedience and direct action on behalf of climate protection were rare. In 2002 Greenpeace launched one of the first actions, a shutdown of Esso gas stations in Luxembourg to protest the company's "continued sabotage of international efforts to protect the climate."[37] In 2005, the Mountain Justice Summer movement disrupted corporate meetings and blockaded mining facilities to halt mountaintop removal mining in Appalachia, culminating in the arrest of climate scientist James Hansen, actress Daryl Hannah, Rainforest Action Network director Mike Brune, and twenty-six other protestors.[38] Starting in 2006, "climate camps," drawing on the experience of the "peace camps" of the 1980s, combined trespassing and squatting with workshops on organic farming and alternative economics. The first of these may have been the ten-day Camp for Climate Action held near the Drax coal-fired power station in West Yorkshire, United Kingdom, in protest of its high carbon emissions.[39] Six hundred activists participated under the watchful eyes of up to 3,000 police, who arrested thirty-eight of the protestors.[40] Such climate camps were subsequently held not only in various locations in Britain but also in Denmark, Canada, France, Ireland, Belgium, Wales, and Australia.[41]

Direct action often took dramatic, high-risk forms. In 2007, for example, thirty protestors entered a coal mine in

South Wales and chained themselves to bulldozers and heavy machinery while their compatriots dropped banners and criticized the government's support for coal. In Czechoslovakia, eleven Greenpeace activists climbed the smokestack of a coal plant to protest emissions. Another action saw ten protestors occupying locations in Tasmania's Styx Valley Forest to halt logging.[42] Loose networks like Rising Tide developed to connect direct actionists. In March 2009 in Washington, DC, 2,500 protestors surrounded the Capitol Coal Plant, which provided electricity to the US Congress, and shut it down for a day, leading Congress to commit to switching the plant off coal.[43] In 2013, a group of thirty Greenpeace activists from eighteen countries who protested drilling at an Arctic oil platform operated by Gazprom were held at gunpoint and charged with piracy, but were eventually released. Direct action and civil disobedience have steadily become more common and more central to the climate protection movement since the turn of the millennium.

The Self-Assembly of 350.org

While climate change affected billions of people around the world and millions of them were concerned about it, there was no vehicle for most of them to do anything about it outside the official governmental process of national legislation and international negotiations—and it was not obvious how there could be one. Then a group of college students and their nature-writer professor invented that vehicle.

At Middlebury College in Middlebury, Vermont, a group of students who were taking a course with economics professor Jonathan Isham started meeting in dining halls and dorm rooms over breakfast and evening beers to discuss how they could take action on global warming. Other students began joining the informal sessions. In February 2005 the group started meeting every Sunday night, dubbing itself the Sunday Night Group; this assemblage participated in a variety of small-scale actions around global warming.[44] Its frame was that "climate change is a defining issue of our time and needs to be addressed."[45] Bill McKibben, a former *New Yorker* writer teaching at Middlebury whose 1989 book *The End of Nature* was perhaps the first book on climate change for a general audience, was an early supporter of the group, regularly attending meetings and speaking at events when asked.

The group had no president, secretary, or treasurer. *Fight Global Warming Now*, a book written by McKibben and six veterans of the Sunday Night Group in 2007, noted that, "To the degree that leaders emerge, it is a result of natural leadership associated with a person's skills or talents around a project, or the time he or she has at the moment, rather than titles or positions."[46]

The End of Nature had urged action to protect the climate, but seven years after its publication McKibben felt that policymakers still were ignoring the evident threat while carbon emissions poured into the atmosphere at an ever-increasing rate. In August 2006 he decided that something had to be done about global warming other than writing books about it. He later recalled that he was "somewhat clueless about spurring activist movements."[47] McKibben's original plan was to walk across Vermont with a few friends and then hold a sit-in on the steps of the state capitol until he was arrested, but someone found out that Burlington police would let them sit there as long as they wanted without arrest. Three weeks later, McKibben and the Sunday Night Group led 300 people on a five-day walk across the state, culminating with 1,000 participants for the final march into Burlington.[48] It was the largest climate demonstration ever held in the United States up to that time.

The Middlebury team developed an experimental testing approach that characterized a series of subsequent initiatives. McKibben and the Sunday Night Group decided to "go national" by using social media to ask people to organize climate rallies around the United States on April 14, 2007. They chose the name Step It Up 2007 to emphasize the urgency of climate protection. Echoing the analysis of the IPCC and the targets and timetables of the Kyoto agreement, the group demanded that Congress enact policies that would cut US carbon emissions by 80 percent by 2050. McKibben told an interviewer that the purpose was "to start a people's movement about climate change. And to shift the debate on Capitol Hill in a more ambitious direction."[49]

The group started with close to zero money and only a vague idea of how the actions it was promoting would come together. The concept was simply to invite diverse people to participate on their own terms. "For three months, our small team was crammed into one room at our headquarters in Burlington, Vermont. Day after day, from this little corner of the country, we peered into our laptop screens and witnessed the self-assembly of a nationwide movement." As new groups organized, "digital pins" populated a "digital map." "Tens of thousands of people went online to post announcements, share ideas, debate tactics, exchange resources, and inspire one another with their passion and creativity." The result was what the organizers described as "a Web-based collaborative effort that allows anyone involved to add their own flavor to the project."[50] People became part of the movement just by organizing an action or showing up for one.

McKibben recalled, "We thought, maybe, we could organize a couple of hundred of these actions. But by mid-February we'd blown by the 650 mark. There are sororities and retirement communities and national environmental groups and churches and rock-climbers and you name it—people were simply waiting for the opening to make their voices heard."[51] On April 14, there were 1,400 actions in all 50 US states, ranging from skiers descending a melting glacier to divers demonstrating climate change effects under the ocean.

That winter, NASA climate scientist James Hansen warned that any carbon dioxide level above 350 ppm was unsafe. "If humanity wishes to preserve a planet similar to that on which civilization developed and to which life on Earth is adapted, paleoclimate evidence and ongoing climate change suggest that CO_2 will need to be reduced from its current 385 ppm to at most 350 ppm, but likely less than that."

The 350-ppm standard transformed climate change discourse and action, which had become preoccupied with increasingly arcane discussions of targets and timetables for various countries and classes of countries. Having a target of "350 ppm" put the spotlight back on what needed to be achieved by all parties acting together. It also redefined the problem as one of common preservation, rather than national advantage. Instead of endlessly debating hypothetical carbon reduction targets for two or four decades in the future, climate protection advocates could zero in on the urgent necessity to start reducing the levels of carbon already in the atmosphere to those scientists said were safe. That changed the game in the global climate debate, as it allowed a focus on the goal to be realized, rather than simply laying out hypothetical policy options. It also defined every reduction in GHG emissions as a contribution to fighting climate change.

The Middlebury team and a small crew of fellow travelers based in countries all over the world, including South Africa, India, and New Zealand, decided to take the model of Step It Up and go global. Naming their new initiative 350.org, they called for global demonstrations on October 24, 2009, shortly before the Copenhagen climate summit. That day saw more than 5,200 actions in 181 countries, from Mt. Everest to the Great Barrier Reef, in what CNN declared "the most widespread day of political action" in the planet's history.[52]

After the Copenhagen fiasco, the climate protection movement was very much at a loss about how to proceed. There was no apparent way to get either nations or world leaders to recommit to genuine climate protection. In response, 350.org continued experimenting with a variety of kinds of action.

For example, 350.org, Greenpeace, and some other groups decided to focus on positive steps to protect the climate through what they dubbed a Global Work Party. Always on the lookout for catchy numbers, they set the date as 10/10/10. Again the groups used social media to reach out around the world—this time with the credibility and vast set of contacts developed during the first Day of Action. On 10/10/10, millions of people around the world installed solar panels, erected wind turbines, and planted community gardens. Participants included "bike repair workshoppers in San Francisco, school insulating teams in London, waste-land-to-veggies-gardeners in New Zealand, and solar panel installers in Kenya."[53]

Concluding that no amount of pressure was going to get government leaders to realize ambitious policy goals in the short run, 350.org started a campaign simply to get world leaders to install solar panels on their own roofs. It encouraged local actions in places where the movement was strong: A petition campaign opposed fracking in the Delaware River basin, for example.[54] In addition, 350.org used global mobilization to support local actions, for example joining an international campaign to force the World Bank not to finance a large coal-fired power plant in Kosovo[55] and mustering support for the India Beyond Coal national day of action.[56]

A year after the 10/10/10 Global Work Party, 350.org initiated a Moving Planet day of action. On September 24, 2011, people at over 2,000 events in more than 175 countries marched, biked, and skated as a way to call for the world to move beyond fossil fuels.[57] A petition campaign coinciding with the Durban COP in December 2011 urged the United States to abandon a plan to delay a climate deal until 2020.[58] Another petition supported global climate leader President Mohamed Nasheed of the

Maldives when he was overthrown by a coup d'état.[59] Immediately after the emergence of Occupy Wall Street, McKibben made a support visit to the encampment, and Occupy activists began joining 350.org climate actions around the country.[60]

Moreover, 350.org experimented with breaking the media wall of silence about climate change. For example, it initiated a "350 Radio Wave" designed to get global airplay for an African rap song about climate change called "People Power." The rap was played in sixty countries, in areas from Texas to Capetown, and went viral on the web.[61] In another operation, 350.org also launched petition drives against some of the most serious opponents of climate protection. The US Chamber of Commerce has been a significant vehicle for the fossil fuel industry to exercise hegemony over the rest of American business on climate issues. As some companies began pulling out of the chamber because of its climate policies, 350.org launched a campaign called The U.S. Chamber Doesn't Speak for Me.[62] When the leading climate denialist, Heartland Institute, posted a billboard with a picture of Unabomber Ted Kaczynski and a text reading "I still believe in Global Warming. Do you?" 350.org and other groups started a petition demanding that the ad be withdrawn, the Heartland Institute apologize, and corporations cease funding its work.[63] Many companies withdrew from both the chamber and the institute.

350.org maintained that what it described as "decentralized networking" is "exactly what we need to fight global warming, that is, a sustained and lively social movement." Rather than "restricting communications to messages flowing up and down (as in traditional hierarchy)," we have to "allow people at the 'ends' of our networks to connect and information to flow in every direction." For the Step It Up day of action, "we made sure that individuals across the nation could connect to each other online in our 'organizer forum,' a chat room and discussion board on the site where people could brainstorm and tackle problems together."[64] (Later they would make their own organizing software available for all comers to adopt.[65])

This decentralization involved a highly fluid form of organization. They advised, "Don't fret about structure. Far more than we need new organizations, we need nimble, relevant, strategic, and often temporary groups of people who can come together to do what needs to be done at the moment—and then do it again, with a whole different bunch of people, a few months later."[66] 350.org would carry its "decentralized networking" into its subsequent campaigns against the Keystone XL pipeline and for divestment from fossil fuel corporations—but use it in a much more sustained way than its one-shot Days of Action.

While the success of efforts like Step It Up and 350.org have often been attributed to the Internet and social networking tools, the organizers emphasized that those tools are not a substitute for face-to-face community. "The actions themselves didn't happen on line—they were real-life, on-the-ground affairs, with neighbors coming together in the flesh to demand change. We feel strongly that the Internet is best used to get people together face-to-face."[67]

Keystone XL Pipeline

Deep under the boreal forest of northern Alberta lie buried the remains of another forest in the form of a viscous petroleum known as bitumen or tar sands. Most of it was deeply buried and regarded as too costly to extract and process until the price of petroleum shot up more than fivefold after 9/11.[68] Oil companies subsequently began extracting and processing the tar sands, taking much of the area's water and leaving miles of despoiled forest and huge lagoons contaminated by waste and chemicals in their wake. Because the tar sands oil was so inaccessible, companies began promoting pipelines to carry it away.

In 2008 the TransCanada Corporation applied for a permit to build the Keystone XL (KXL) pipeline, which would carry 800,000 barrels of Alberta bitumen daily 1,700 miles across the US border to refineries in Texas.

The Alberta tar sands region was the home to Cree, Dene, Métis, and other indigenous First Nations. While some welcomed the jobs and economic opportunities that came with the tar sands extraction, many were vehemently opposed, especially as the devastation to land and water—and the consequent destruction of traditional subsistence hunting and fishing—became increasingly blatant and residents began to exhibit high rates of cancer and autoimmune diseases. Eventually, tribes opposed to the mining joined with dozens of other organizations across Alberta to call for a moratorium on tar sands projects, filing suit to halt specific projects.[69] Intertribal gatherings and the Indigenous Environmental Network (IEN) began constructing anti-KXL alliances with other indigenous groups around North America.

The KXL pipeline was slated to pass through Nebraska's famous Sand Hills and over the Ogallala Aquifer, which if contaminated would pollute the water supply for much of the Great Plains. A new organization called BOLD Nebraska pulled together a coalition of conservative ranchers, urban progressives, environmentalists, and farm advocacy groups to oppose the pipeline.[70]

Initially the opposition to tar sands extraction and the KXL pipeline, like many local campaigns against the extraction of extreme energy, grew out of local environmental and health concerns and attracted little attention beyond the affected local areas. Then James Hansen, head of the NASA Goddard Institute and widely regarded as the leading climate scientist in the United States, wrote in his blog that tar sands "must be left in the ground." Indeed, "if the tar sands are thrown into the mix it is essentially game over" for a viable planet. That statement defined tar sands extraction not just as a local issue, but rather as a key element of the global climate crisis.

One person who took note of Hansen's blog was Bill McKibben, who had recently taught a Middlebury College course on social movements that had read Taylor Branch's three-volume study of Martin Luther King, Jr., and the civil rights movement. McKibben had already concluded that it was time for the climate movement to turn to civil disobedience. What particularly struck him about the KXL pipeline was not only that it threatened the planet, but that because it crossed a US border it required approval by the president of the United States, who could deny a permit for it without action by the climate-deadlocked US Congress. In June 2011, after checking with the original pipeline opponents in Alberta and Nebraska, McKibben and a group of other movement leaders sent out a letter calling for a month of civil disobedience actions at the White House to demand that President Obama deny the KXL pipeline a permit.

Although the effort was connected to 350.org, it was organized by a new entity, Tar Sands Action, "so we could have a broad front supported by many groups."[71] McKibben and other initiators were arrested the first day of the protest and locked in jail for several days. Those who arrived in Washington from around the country were met not by organization leaders but by trainers who briefed them on legal and logistical conditions and helped them role-play their sit-in and arrest. Over a two-week period more than 1,200 people were arrested in the ongoing protests. McKibben concluded that "establishment, insider environmentalism found itself a little overtaken by grassroots power"; the movement had "gone beyond education to resistance."[72]

The White House sit-ins led to a cascade of consequences. Ten large environmental organizations sent a joint letter to President Obama supporting the demonstrations and opposing KXL.[73] Ten Nobel Peace laureates, including Desmond Tutu and the Dalai Lama, asked Obama to block the pipeline.[74] A million comments flooded the State Department comment line.

The campaign against the KXL pipeline also brought divisions. Back in 2010, TransCanada had signed a coveted "project labor agreement" with the Teamsters, Plumbers, Operating Engineers, and Laborers unions offering preferential hiring and union conditions on the KXL pipeline. The four union presidents issued a statement saying that the project would "pave a path to better days and raise the standard of living for working men and women in the construction, manufacturing, and transportation industries."

The media rushed to portray the pipeline as a typical case of "jobs versus the environment" pitting the labor movement against the environmental movement; NPR's headline was "Pipeline Decision Pits Jobs against Environment." There was also opposition to the pipeline within organized labor, however. The Transport Workers Union (TWU) and the Amalgamated Transit Union (ATU) issued a joint statement saying, "We need jobs, but not ones based on increasing our reliance on Tar Sands Oil." The statement called for major New Deal–type public investments in infrastructure modernization and repair, energy conservation, and climate protection as a means of "putting people to work and laying the foundations of a green and sustainable economic future for the United States." Terry O'Sullivan, president of the Laborers' Union (LIUNA), cracked back: "It's time for ATU and TWU to come out from under the skirts of delusional environmental groups which stand in the way of creating good, much-needed American jobs." The divided AFL-CIO first maintained public neutrality on the pipeline, then gave it tacit support, and finally publicly lobbied Obama to authorize it.[75]

In November 2011, 15,000 people surrounded the White House with a mile-and-a-half-long "solidarity hug." A presidential veto of the pipeline seemed a very long shot, however: A *National Journal* poll of three hundred energy insiders found that more than 90 percent of them believed the KXL would get its permit.[76] Four days after the "solidarity hug," the White House announced that the State Department needed another year to study the issue.

In December 2011, Republicans in Congress tried to force President Obama's hand by submitting legislation requiring a decision within sixty days. Major environmental organizations and other tar sands opponents conducted an inside-the-Beltway lobbying campaign against it, backed up by a grassroots mobilization. The legislation passed the House, 234 to 193; McKibben estimated that the fossil fuel industry had spent $42 million to win the vote.[77] The powerful American Petroleum Institute said that there would be "huge political consequences" if the pipeline was blocked.[78] However, it was defeated in the Senate by two votes.[79]

The KXL pipeline had become a leading national political issue. Mitt Romney's first TV ad of the general election campaign began:

> VOICE-OVER: "What would a Romney presidency be like?"
> VIDEO TEXT: "Day 1"
> VOICE-OVER: "Day one, President Romney immediately approves the Keystone pipeline, creating thousands of jobs that Obama blocked."[80]

The *New Yorker* observed that KXL had become "the most prominent environmental cause in America."[81] McKibben reflected, "We managed to put something no one knew about at the center of the nation's political agenda."

The KXL campaign represented a broader transformation in environmental politics. The Sierra Club, which for 120 years had advocated using only "lawful means," for the first time endorsed civil disobedience. In February 2013, the Sierra Club, with the unified support of the large environmental groups, NAACP president emeritus Julian Bond, and the Hip Hop Caucus, led nonviolent civil disobedience at the White House, followed by the largest climate rally in US history, with an estimated 40,000 participants.[82]

A 2013 article by Katherine Bagley of *InsideClimate News* detailed the transformation in environmental politics wrought by the KXL campaign. In 2009, the "main engine" of climate protection were the biggest green groups, which "spent vast quantities of financial and political capital" lobbying for climate legislation. By 2013, "the main force" was "a messy amalgam of disparate grassroots efforts stretching from Maine to Utah" that has "found common cause" in opposing the KXL pipeline and other tar sands projects. One week in May saw "hundreds of activists march at an Obama fundraiser event" in New York City; two citizens "encase themselves in concrete at a construction site" for the southern leg of the Keystone; citizens of Utah "flood the email inboxes and fax machines of investors for a proposed tar sands mine in Utah with the message 'We will stop you before it starts'"; and hundreds protest the appearance of Canadian prime minister Stephen Harper at the Council on Foreign Relations headquarters in Manhattan.[83]

When President Obama authorized construction of the southern leg of the pipeline, environmental activists and threatened landowners formed a loose resistance movement known as Tar Sands Blockade. An article in *Rolling Stone* detailed how members "locked their necks to excavators in Oklahoma, sealed themselves inside pipe segments in Winona, Texas, and stormed TransCanada's Houston offices." Activists staged a three-month "tree sit-in" in east Texas that forced TransCanada to reroute the pipeline around the blockade.[84]

The anti-KXL movement also began to converge with other on-the-ground direct action campaigns. A February 2013 "anti-extraction summit" of environmental justice organizations initiated #FearlessSummer to organize direct action in "front-line communities" most directly affected by energy extraction and processing. In cooperation, 350.org organized Summer Heat to mobilize its constituency to support such actions. Results included more than two hundred arrests at a Chevron oil refinery in Richmond, California; forty-five arrests at a coal-fired power plant outside of Boston; a flotilla of kayaks protesting a natural gas storage facility on Seneca Lake in New York; and other direct actions in Utah, Texas, and West Virginia.[85]

Opposition to tar sands extraction and the KXL pipeline had started with indigenous people, and it spread far and wide through Indian country. The KXL route crossed hundreds of sacred sites and threatened major sources of drinking water for large reservations. The burgeoning Canadian First Nations movement Idle No More, which was founded to challenge treaty-rights violations, took on KXL as a central issue. Tribal councils from the Lakota to the Nez Perce passed resolutions opposing the pipeline. In January 2013, leaders from many Indian nations journeyed to Pickstown, South Dakota, to draft a document rejecting the KXL and the rest of the tar sands infrastructure. A Training for Resistance tour included a Moccasins on the Ground gathering on the Pine Ridge reservation in South Dakota, preparing for direct action should the pipeline be approved.[86]

Seeking a preemptive deterrent to pipeline approval, the political action–oriented telephone company CREDO joined with the Rainforest Action Network and The Other 98% to start the Keystone XL Pledge of Resistance. The campaign was modeled on the 1984 Pledge of Resistance, in which 100,000 people promised to commit civil disobedience if the United States invaded Nicaragua; both pledges aimed to deter a political decision by raising its political cost. By May 2014, more than 95,000 people had pledged to risk arrest if the State Department gave the go-ahead to the pipeline.[87] To show the Obama administration that the threat was real, the KXL Pledge of Resistance trained hundreds of local action leaders, helped them develop local action plans in nearly two hundred communities, and organized small "warning-shot" civil disobedience actions in Chicago, Houston, and Boston.[88]

In late April 2014, an improbable but highly photogenic coalition that called itself the Cowboy Indian Alliance of farmers, ranchers, and tribal leaders from the pipeline corridor held a five-day Reject and Protect encampment in Washington. A week before they were scheduled to arrive, the Obama administration announced that its decision on the pipeline would again be delayed, probably for many months.

The KXL struggle brought echoes from American crises past. Reverend Lennox Yearwood of the Hip Hop Caucus said, for the civil rights movement, "Birmingham became the epitome." The KXL "became in essence our Birmingham ... the Birmingham for the environmental movement." Defeating the pipeline "will have a ripple effect on the other parts of the country and the world" comparable to successes of the civil rights movement in Birmingham. [89] If Obama approves the KXL pipeline, said Michael Brune, referring to the fate of Lyndon Johnson, it will be "the Vietnam of his presidency."[90]

The KXL campaign grew out of the inability to address climate protection through normal legislative and electoral means, embodied in the apparent impossibility of passing climate legislation through the US Congress. It also reflected the climate movement's need to move beyond one-day events to ongoing dramatic confrontations that would engage with the power of its opponents and project climate protection into the public mind and the political arena.

The campaign brought together local campaigns against local environmental destruction along the pipeline's route, climate activists in 350.org and beyond, environmental justice groups like the Hip-Hop Caucus, direct actionists like

Tar Sands Blockade, and large environmental organizations like the Sierra Club that had never previously supported civil disobedience.

If the KXL pipeline is ultimately approved, it will be a painful defeat for the climate protection movement, but it will also be a stunning example of winning by losing. The campaign against the KXL has educated millions of people about climate change, drawn diverse elements of the climate protection movement into shared goals and coordinated action, and introduced forms of action that provide the movement with new possibilities of power. Win or lose, the climate protection movement will emerge from the KXL fight far different and far stronger than it was before.

Divest-Invest

In 2010, a group of students from Swarthmore College visited mountaintop-removal mining operations in West Virginia. One participant recalled, "It's devastating. It's people's land, people's livelihoods being ripped out of the ground, literally." The students initiated a campaign to get Swarthmore and other colleges to withdraw their investments from coal companies that conducted mountaintop-removal mining.[91] In June 2011, a meeting of student groups from several colleges organized a Divest Coal campaign, which spread to fifty campuses within a year.[92]

Meanwhile, the campaign against the KXL pipeline was being criticized for focusing on one tiny part of the climate problem without articulating a broader vision of what needed to be done and how to make it happen. Bill McKibben understood the force of the argument, recognizing there was "no way we could slow global warming one pipeline at a time." Instead, he concluded, it was necessary to "challenge the underlying legitimacy of the whole coal and gas and oil machine."[93]

McKibben wrote a widely discussed article for *Rolling Stone* called "Global Warming's Terrifying New Math." It warned that, according to current climate science, people can put no more than 565 gigatons of carbon into the atmosphere without warming it beyond the 2°C danger line. The proven coal, oil, and gas reserves of the private and government-owned fossil fuel companies would, if burned, release 2,795 gigatons of carbon into the atmosphere—five times as much as needed to drive global warming over 2°C. Those reserves formed the assets of the fossil fuel companies, worth about $28 trillion at current prices, and the companies were still searching avidly for more. The business plan of the fossil fuel companies was, in essence, to destroy the climate.[94]

Taking a leaf from the student campaign for coal divestment already under way and harkening back to the historic campaign to withdraw investment from apartheid-era South Africa, the article called for divestment from all fossil fuel companies. McKibben and 350.org followed up with a twenty-city Do the Math road show calling for individuals and institutions to get rid of their fossil fuel investments. Within ten days, students had organized "fossil free" divestment campaigns on two hundred campuses.[95] In late November, 72 percent of Harvard University undergraduates participating in student elections voted for the university to divest from fossil fuel companies.[96] Similarly, students at colleges around the country began holding meetings with college officials and trustees demanding divestment. The *New York Times* said the students were at the "vanguard of a new national movement."[97] In May 2014 Stanford University agreed to eliminate its holdings in one hundred coal companies in response to student demands, the first major US university to start divesting from fossil fuels.[98]

According to a 2013 study of the divestment movement by University of Oxford's Smith School of Enterprise and the Environment, six colleges and universities had already committed to divest, along with seventeen cities, two counties, eleven religious institutions, three foundations, and two other institutions.[99] The Smith School study compared the campaign with five previous divestment campaigns and concluded that it was growing faster than any of them, also noting that the campaign could cause significant damage to coal, oil, and gas companies. While the direct financial effect of disinvestment by university endowments and public pension funds would be relatively small, the study found that "the outcome of the stigmatisation process, which the fossil fuel divestment campaign has now triggered, poses the most far-reaching threat to fossil fuel companies and the vast energy value chain." Report coauthor Ben Caldecott said, "Stigmatisation poses a far-reaching threat to fossil fuel companies." [100]

In January 2014, seventeen philanthropic foundations with combined assets of $2 billion announced that they had decided to divest from fossil fuels and invest in clean energy. Ellen Dorsey and Richard Mott of the Wallace Global Fund, which helped pull together the initiative, observed that "by building a movement that defines the fossil fuel industry as a moral pariah, Divest-Invest seeks to break the industry's grip on our political process." They point out that there may also be a dimension of economic self-interest in divestment, writing that "fossil fuel stocks may be massively overvalued" because restrictions on GHG emissions may make them unusable "stranded assets." The "Invest side" of the campaign focuses on "climate solutions" and "the need for massive increase in clean tech investments." They envision expanded investment in "green loan funds," "green bonds," "green banks," and "energy projects that transform our local economies."[101] By September 2014, 180 institutions— including the Rockefeller Brothers Fund, whose wealth derived from Standard Oil—had pledged to sell fossil fuel assets worth more than $50 billion.[102]

Notwithstanding its promise, there are some problems with fossil fuel divestment. The antiapartheid divestment campaign produced an interaction of stigmatization and direct financial harm to the fragile South African economy; while divestment can effectively stigmatize the fossil fuel industry, few claim it will directly harm the industry's bottom line. It is clear that much of the companies' fossil fuel reserves must not be burned if climate disaster is to be contained; however, it is not clear that forcing companies to leave some deposits in the ground without an overall global regulatory regime won't just raise prices for the rest, leading to greater extraction elsewhere. While there has been a call to invest funds that have been withdrawn from fossil fuel companies in energy alternatives, so far there are few concrete examples. Finally, the focus on fossil fuel companies, while necessary, does not address the wider constellation of social forces and structures—ranging from governments to energy consumers—that help perpetuate their destructive role.

★   ★   ★

The failure of established institutions to solve problems has often led to the emergence of movements demanding radical change and using more radical methods. Betrayed government promises for racial equality and nuclear disarmament, for example, helped spawn national and global civil rights, ban-the-bomb, and student movements in the 1960s. Climate protection from below has similarly emerged from the failure of climate protection from above.

Climate protection from below has evolved rapidly and has formed a global movement. It has grown independent of any nation-state and of any corporate interest, willing to challenge both. It has established a networked form of organization that is highly flexible while facilitating rapid coordination and mass mobilization on a global scale, drawing tens of millions of people into grassroots self-organization. Climate protection from below has established a common interpretive frame—establishing climate protection as a common objective—and has related that frame to issues of social justice. It has projected its frame and objective to hundreds of millions of people and has moved beyond the limits of lobbying to mass civil disobedience.

While the movement has accomplished much, it is still far from reversing the momentum toward climate catastrophe. In plotting the movement's future course, we need to take full measure of the obstacles that have so far foiled all efforts at climate protection. We then need to think about how the movement might develop, drawing on what it has already accomplished, to overcome those obstacles.

Chapter 4
What Climate Protectors Have Accomplished

However limited the efforts at climate protection have been so far, they present a series of experiments from which we can learn many of the elements we will need to succeed in the future. Past efforts provide a foundation, which we will have to modify in the future, but upon which we can build.

Climate scientists have played an essential role in the emergence of climate protection. From the discovery of global climate change to today they have been willing to pursue a subject that almost everyone else seemed determined to ignore. Initially, climate scientists took responsibility for calling global warming to the attention of governments and the United Nations. Many expanded their concept of their responsibility to include alerting the public and eventually taking a leadership role in the climate protection movement—even to the point of going to jail. The scientists' continued role will be a necessary condition for any future climate protection movement.

However hypocritical, and whatever its limitations, the official climate protection effort of concerned governments and the UN represents an acknowledgment of a responsibility to protect the earth's climate. It thereby lays an essential basis for future climate protection efforts. The pursuit of a global agreement with targets and timetables for GHG reduction will remain necessary in the future, even if the mechanisms for implementing it will need to be very different. The basic concept of common but differentiated responsibilities among different nations will also remain essential, even if the means for realizing it will need to change.

The major environmental organizations—sometimes referred to as the "big greens"—took on the climate issue when no one else was doing so. Working through the Climate Action Network they organized globally, supported global efforts at climate protection, and pressured individual countries to pursue long-term common global interests rather than their own short-term advantage. These groups have moved beyond many initial limitations, such as a predominant focus on conventional politics and lobbying, an alignment with developed-country governments, and a reluctance to question the mechanisms proposed for climate protection by governments and conventional economists.

The climate justice movement has played a critical role in expanding climate protection beyond its original rather narrow base to include the concerns and cultures of many groups initially left out by both official and civil society efforts. It has also provided an important critique of both the efficacy and the fairness of the mechanisms proposed for climate protection. Because climate change and policy to address it have a huge differential impact on different groups, issues of justice will remain inescapably central to the future of climate protection.

Those who have pursued climate protection legislation and policies in national political arenas have set much of the groundwork for future climate action, especially in those countries that have taken significant steps forward. While the specific policies they have pursued and the compromises they have made will need to be revised in order to effectively protect the climate, the need to use the power of national governments to enforce rapid GHG emission reductions and provide energy and economic alternatives grows more evident every day.

Local efforts by municipalities and civil society groups are also laying the groundwork that will be needed in the future. Decentralized initiatives with wide public participation will be crucial for an effective transformation to a climate-safe economy and society. Only experimentation at the local level can provide the detailed knowledge of what can really work for that transformation.

The emergence of a direct action movement using the tactics and traditions of civil disobedience marked a critical advance in the development of climate protection as its methods moved beyond the limitations of electoral and lobbying activity. It challenged individuals and institutions to stop disregarding the "inconvenient truth" of climate change. In the future it will be necessary as a means to challenge the power and legitimacy of the fossil fuel–producing and –using industries and the governments that protect them.

The global days of action initiated by 350.org created a way for people around the world to participate together in demanding protection of the climate. The grassroots self-organization, decentralized networking, and combination of virtual communication and in-the-flesh organizing and action will be a necessary basis for a climate movement that can transform the world.

The campaign against the KXL pipeline represents many of the elements that will be essential for a successful climate protection movement in the future. It has used civil disobedience and direct action, combining them with public education and lobbying when necessary. This campaign has drawn together forces that reflect the full spectrum of people affected by climate change.

The fossil fuel divestment movement has begun the long, hard job of stigmatizing the fossil fuel industry, changing its image from a friendly purveyor of wonderful resources to a destroyer of the earth and the future of its people. While divestment is unlikely to be the tactic that ultimately leads to a transformation of our energy system, the mobilization against the purveyors of destruction that it represents will be central to effective climate protection.

The weekend of September 21, 2014, half a million people joined 2,646 events in 162 countries to demand global reductions in greenhouse gas emissions. An estimated 40,000 marched in London; 30,000 in Melbourne; 25,000 in Paris. Some 400,000 joined the Peoples Climate March through the center of New York City. The climate protection movement had come a long way since 2006, when a march of 1,000 through Burlington, Vermont, proved to be the largest climate protest in American history, or since 2013, when a 40,000-strong protest was the largest US climate demonstration up to that time. Yet despite its exponential growth, whether and how the climate protection movement could realize its goals remained an open question.[1]

To be effective, any future climate protection movement will have to go far beyond, both in size and power, anything that has come so far. I believe that the movement will have to become, in effect, a global nonviolent insurgency in which hundreds of millions of people challenge the power and legitimacy not only of the fossil fuel industry but also of the corporations and governments that provide support for its power. If such a movement does in fact emerge, and if it imposes the changes that are necessary to make humanity climate-safe, future historians will find its antecedents in the climate protection efforts of the past twenty-five years.

CHAPTER 5
WHY CLIMATE PROTECTION HAS FAILED

Scientists and climate protection advocates once expected that rational leaders and institutions would respond appropriately to the common threat of climate change. As Bill McKibben said of James Hansen and himself, "I think he thought, as did I, If we get this set of facts out in front of everybody, they're so powerful—overwhelming—that people will do what needs to be done."[1]

What went wrong? Why has the world's obvious long-term common interest been so hard to realize?

The disturbing answer is that the measures we need to protect the global ecosphere threaten the power of the world's most powerful institutions. National governments will have to accept international controls, while corporations will have to give up opportunities to make money at the expense of the environment. Military establishments will have to abandon weapons and wars that threaten the air and water. Beyond that, virtually everyone will have to adjust to a substantial change—though not necessarily deterioration—in lifestyle.

Governments, corporations, and other dominant institutions are not evolved to provide for either the long-term interests or the common interests of the world's people. These dominant institutions have grown and prospered by pursuing the short-term interests of their citizens and stockholders (or often just a small, dominant elite among them) in competition with the citizens and stockholders of other companies and countries. They are not designed or structured to pursue any wider human or global interest. Moreover, their time horizon is determined not by the lifetimes of our children and grandchildren but by the next election cycle or quarterly report. To their leaders, sustainability means getting through the next couple of years without loss of elections or profits.

Climate protection advocates had erroneous expectations because these institutions and leaders were willing to give lip service to climate protection, and even use its advocacy to advance their own competitive position. When it came to actually doing what was necessary to protect the global climate, however, the institutions' own short-term national and corporate interests came first.

Conversely, the institutions supposed to represent global common interests—for example, the UN—proved weak and dependent on governments, which ultimately retain formal or de facto veto power over their actions. Even the IPCC—ostensibly a scientific organization—is made up overwhelmingly of government-employed scientists, has its reports reviewed by government officials, and requires the wording of its influential *Summary for Policymakers* to be approved line-by-line by all of the more than 120 participating governments. Most governments, in turn, are subject to the de facto veto power of private economic interests driven to pursue short-term private gain above all else.

While great powers and corporations are the dominant factors in this process, many other people and institutions pursue short-term self-interest at the expense of climate protection, often in pursuit of their own economic survival. Local communities and workers dependent on fossil fuel industries, for example, have campaigned to weaken climate protection legislation and block international climate agreements. Developing countries have fought to maintain their right to expand their use of coal. Such de facto allies have helped enable the major GHG emitters and their supporters to pursue a hypocritical path, talking the climate protection talk while walking the GHG walk.

World Order Obstacles to Climate Protection

Climate destruction is not the result of action by people whose aim is to destroy the climate. Rather, it results from people operating within institutional structures in which they pursue goals and practices the effect of which—whether they know it or not—is climate destruction. Such structures include:

Fossil Fuel–Producing Industry

The most obvious purveyor of climate destruction is the fossil fuel industry. Just ninety private and government-owned companies, nearly all of them fossil fuel companies, are responsible for nearly two-thirds of all carbon emissions since 1750—with half of them produced in the past twenty-five years.[2] Nineteen of the world's fifty leading corporations are fossil fuel–producing companies and utilities. They account for 48 percent of the revenues and nearly 46 percent of the profits of the top fifty companies in the Fortune Global 500.[3] Climate protection means abolishing the fossil fuel industry as we know it and rendering its primary asset—fossil fuels—worthless. The industry understands that and spends billions of dollars to corrupt politicians, dominate elections, and brainwash the public. The industry also knows that at present almost all human purposes depend on fossil fuels; it uses that dependence to wield hegemony over nations, peoples, and institutions. In addition, it implicitly and explicitly threatens that if it doesn't get its way we will all end up shivering in the dark.

Network of Support for Fossil Fuels

Theorist of nonviolence Gene Sharp has argued that the power of the powerful depends on "pillars of support"—institutions and constituencies that supply them with the needed sources of power.[4] Surrounding the fossil fuel–producing industry is a wide swath of forces that advocate for its interests; they are often interpenetrated with it, dominated by it, and dependent on it. Supporters include fossil fuel–using industries, the financial industry, anti–climate protection corporations, politicians and political parties, much of organized labor, and people and institutions who believe they are dependent on fossil fuels to meet their daily needs.

Neoliberalism

Neoliberalism is an ideology that argues that global market forces should determine human decisions and that governments and other public institutions should act only to support private profitmaking. Neoliberal ideology is not just a theory propounded by economists; it guides the action of the dominant economic institutions of the world order, including the major banks, corporations, the treasury departments of the United States and other countries, the IMF, the World Bank, and the WTO. The ideology is used to argue against taxes, regulation, public investment, and other use of public authority for any purposes except promoting private profit and to oppose "interference" with corporations doing whatever they decide—including destroying the earth's atmosphere. Neoliberalism thereby plays a crucial role in preventing effective climate protection.

Nation-State System

Under the established doctrine of nation-state sovereignty the government of each nation is legally authorized to decide its actions without interference. Under this theory, no larger or longer-term interest can be imposed on nations except by their own consent. This doctrine, embodied in the practice of states and the structure of the United Nations, has allowed nations to lay waste to the atmosphere and the common future of humanity. Further, a competitive nation-state system generates conflict and antagonism among nation-states, as it drives states to pursue advantage in the competition with other states. Accumulation of economic power is a crucial form of this competition, driving expansion of gross domestic product (GDP) even where the consequences for a country's own citizens, let alone the rest of the world and future generations, are horrendous. Finally, this system allows corporations and other private actors to pursue their destruction of the earth's atmosphere behind the shield of national sovereignty.

Within this system, effective power is largely concentrated in a few dominant nations, often leading coalitions of other countries. While the United States and its allies dominated this system during the twentieth century, their hegemony is now being challenged by China and other rapidly developing nations. The result at present is a de facto alliance of the largest GHG emitters, led by the United States and China, who starting with Copenhagen have cooperated to defeat efforts at effective climate protection.

Dependence on fossil fuels, neoliberalism, the nation-state system, and the great power struggle for hegemony are not primarily features of one or another nation. Rather, they are properties of the world order—the overall patterns by which our species has organized its life on earth. As international law scholar Richard Falk put it on the eve of the Copenhagen climate summit, the inability of governments to cooperate to protect global public interests is compounded of "statism, neoliberal capitalism, hegemonic geopolitics, presentism, militarism, and nationalism."[5]

Obstacles in Human Hearts and Minds

In addition to these institutional, structural obstacles, there are obstacles that impede the coming together of individuals and social groups to take collective action to halt climate change, even though to do so is in their individual and collective interest. They include:

Denialism

Denial of global warming can take the form of directly rejecting the science and evidence supporting it, based on pseudoscience or ignorance. It can, however, also take the form of simply ignoring it or paying much more attention to other things. Denial has been deliberately promoted by the fossil fuel industry and its allies and supporters (including the political right wing in the United States) as a weapon against climate protection policies—taking a leaf from the tobacco industry's long denial of the health effects of smoking. Nearly all of us repress our consciousness of the climate threat some of the time, trying to tune out something that is too overwhelming to contemplate without losing our equilibrium and becoming unable to go on with our lives.

Incrementalism

Many people, including many politicians and leaders of businesses, unions, and other institutions, admit the reality of climate change but do not support the kinds of "extreme" measures necessary to halt it now. Some downplay the significance of climate change as a universal, existential threat and a clear and present danger not just to polar bears but to humanity. Some people say it is not politically realistic to address it aggressively, that we need to start slowly and put off serious reductions in GHGs until long in the future. There is even such misleading incrementalism within the climate movement itself, taking the form of an unfounded optimism that inadequate but politically acceptable solutions will suffice.[6]

Economic Consequences of Climate Protection

Many people believe that serious efforts to protect the climate will lead to economic catastrophe for themselves and/or society as a whole. They may feel their jobs depend on the production and use of affordable fossil fuels and may believe that restrictions on or higher prices for fossil fuels will lead to unemployment and economic crisis. Some may fear that climate protection will lead to a sharp rise in prices to consumers. Such fears are fed by a powerful propaganda machine promoting the idea that environmental protection is a threat to prosperity and a "job killer."

Let Another Country Pay

Global climate protection would be in the interest of almost everyone, far outweighing its collective costs, but the system of sovereign nation-states generates a struggle for each country to shift the cost of protecting the climate onto others so as to get the benefits without the costs. Any nation that invests in cutting its own emissions pays the cost, but the benefit is shared among all countries, including those that continue emitting GHGs at breakneck speed.[7] International climate negotiations have come to grief over how the costs and benefits of climate protection should be distributed, while opposition to climate protection in domestic politics often focuses on the demand that "other countries" cut their emissions first. Since the wealth and power of countries and their past, present, and likely future contribution to climate change vary so widely, there is often conflict over the just distribution of the costs of climate protection—resulting in no climate protection at all.

Legitimacy of the Status Quo

In most situations, people accept the legitimacy of the authorities under which they live. Even if they perceive policies like those leading to climate destruction as detrimental to their interests, they don't normally challenge the right of the established authorities to pursue them and to punish those who attempt to interfere with them.

Fear of Social Movements

While some people are thrilled by popular upheavals, many are frightened when they see media images of crowds clashing with police, Molotov cocktails flying through the air, and economies in a tailspin in the wake of social upheaval. Many people also observe that in the aftermath of social upheaval, ordinary people often have less freedom and worse economic conditions than they had before. Some people fear the consequences of social movements, both for their personal well-being and for their society as a whole.

Individualism

The belief in individual rather than collective action forms a barrier to all kinds of social movements. It may take the form of a fear that participation will restrict individual liberty, the belief that people can and/or should look out for themselves, or the conviction that looking out for oneself is likely to be more beneficial than social action in the short and/or long run.

Hopelessness

It is easy to despair that there is anything we can do about climate change, simply because the problem is so devastating and the obstacles to fixing it seem so insurmountable. Even many of us who are devoting our lives to climate protection feel a deep despair about forestalling climate catastrophe. Our efforts seem too little and too late.

*   *   *

The strategy proposed in *Climate Insurgency* is designed to address these obstacles.

Part II

A Plausible Strategy
for Climate Protection

CHAPTER 6

A GLOBAL NONVIOLENT CONSTITUTIONAL INSURGENCY

The global climate protection movement has laid the groundwork for countering the underlying obstacles to climate protection. It has constructed a flexible network form of organization that can facilitate rapid coordination and mass mobilization on a global scale and has drawn tens of millions of people into grassroots self-organization. It has established its independence of any nation-state and of any corporate interest. The movement has established a common interpretive frame and a common objective: the reduction of atmospheric carbon to a climate-safe level, currently estimated at 350 ppm or less. It has related that frame to issues of social justice and has projected its frame and objective to hundreds of millions of people. It has moved beyond the limits of lobbying to mass civil disobedience. The global climate movement has become one of the power actors of the world order, able to challenge states, corporations, and other central institutions.

To realize its objectives, the climate protection movement must now use the capacities it has created to overcome the obstacles to climate protection presented by the organization of our current world order. The movement must limit the blind pursuit of self-interest and self-aggrandizement by states and corporations and must nurture means for formulating and pursuing the global common interest in protecting the climate. Moreover, it must overcome the GHG-protecting hegemony imposed by the great powers, above all by the United States. It must develop a strategy for political, economic, and social transformation that protects the climate while protecting people's livelihoods and well-being. Such a strategy doesn't require transforming the world order into some kind of global utopia, but it does mean changing the world order sufficiently to allow effective climate protection. Such a result is plausible only if climate change impacts continue to worsen, the official climate protection process continues to fail, popular concern continues to mount, and the global climate protection movement continues to grow and develop.

Why a Nonviolent Insurgency?

Insurgencies are social movements, but movements of a special type: They reject current rulers' claims to legitimate authority. Insurgencies often develop from movements that initially make no such challenge to established authority, but eventually conclude that it is necessary to realize their objectives. To effectively protect the earth's climate and the future of our species, the climate protection movement may well have to become such an insurgency.

The term "insurgency" is generally associated with an armed rebellion against an established government, one that rejects and resists the authority of the state. Its aim may be to overthrow the existing government, but it may rather aim to change it or simply to protect people against it. Whatever its means and ends, the hallmark of an insurgency is to deny that established state authority is legitimate and to assert that its own actions are.[1]

A nonviolent insurgency pursues similar objectives by different means. Like an armed insurgency, it does not accept the limits on its action imposed by the powers that be. Unlike an armed insurgency, it eschews violence and instead expresses power by mobilizing people for various forms of nonviolent mass action.

After closely following the massive strikes, street battles, peasant revolts, and military mutinies of the Russian Revolution of 1905 that forced Czar Nicholas II to grant a constitution, Mohandas (not yet dubbed "Mahatma") Gandhi concluded, "Even the most powerful cannot rule without the cooperation of the ruled."[2] Shortly thereafter he launched his first civil disobedience campaign, proclaiming, "We too can resort to the Russian remedy against tyranny."

The powers that are responsible for climate change could not rule for a day without the acquiescence of those whose lives and future they are destroying. They are only able to continue their destructive course because others enable or acquiesce in it. It is the activity of people—going to work, paying taxes, buying products, obeying government officials, staying off private property—that continually recreates the power of the powerful. A nonviolent climate insurgency can be powerful if it withdraws that cooperation from the powers that be. Fear of such withdrawal can motivate those in positions of power to change.

Why a Constitutional Insurgency?

It is often pointed out that electoral politics, lobbying, and similar forms of "legitimate" political action accept the established "rules of the game" and operate within their limits. Even if the rules are rigged, participants must accept the outcome of any given round and resign themselves to simply trying again.

The climate protection movement, by adopting civil disobedience, has moved beyond conventional political and lobbying "pressure group" activity to become a protest movement prepared to violate the law. Civil disobedience, while generally recognizing the legitimacy of the law, refuses to obey it. Civil disobedience represents moral protest, but it does not in itself challenge the legal validity of the government or other institutions against which it is directed. Rather, it claims that the obligation to oppose their immoral actions—whether discriminating against a class of people or conducting an immoral war or destroying the climate—is more binding on individuals than the normal duty to obey the law.

A constitutional insurgency goes a step further. It declares a set of laws and policies themselves illegal and sets out to establish law through nonviolent self-help. It is not formally a revolutionary insurgency because it does not challenge the legitimacy of the fundamental law; rather, it claims that current officials are in violation of the very laws that they themselves claim provide the justification for their authority. Such insurgents view those whom they are disobeying as merely persons claiming to represent legitimate authority, but who are themselves violating the law under color of law—on the false pretense of legal authority. Their "civil disobedience" is actually obedience to law, even a form of law enforcement.[3]

Social movements that disobey established authorities often draw strength from the claim that their actions are not only moral, but that they also represent an effort to enforce fundamental legal and constitutional principles that the authorities they are disobeying are in fact flouting. Such legal justifications strengthen participants by clarifying that they are not just promoting personal policy preferences by criminal means but rather performing a legal duty. Moreover, they strengthen a movement's appeal to the public by presenting its action not as wanton lawbreaking but as an effort to rehabilitate governments and institutions that are themselves in violation of the law.[4]

Existing legal, political, and economic arrangements seem to support the right of those who are conducting and permitting the destruction of the climate to go on doing so. Their authority to do what they do appears legitimate, even though it is leading to species catastrophe.[5]

However, there exist powerful legal arguments that governments are in violation of their most fundamental legal and constitutional duties as long as they permit the destruction of the world's atmosphere. In the next chapter we will explore one of the most promising, the legal principle known in the United States as the "public trust doctrine." The public trust doctrine provides a basis for maintaining that the destruction of the earth's atmosphere, and the collusion of all governments in it, is illegal. It thereby provides one possible basis of legitimacy for a global movement that rejects the claims to legitimate authority of existing governments. No doubt there are others, ranging from fundamental human rights to national and international environmental law.[6] Many of the proposals made in this book for utilizing the public trust doctrine could draw on them as well.

Historical Roots of Constitutional Insurgency

The significance of the public trust principle and other legal claims for climate protection is that they define those claiming authority as illegitimate usurpers so long as they persistently fail to fulfill their duty to protect the atmosphere. The history of social movements shows that such legitimations can play an important role in making change.

For the civil rights movement, the US Constitution's guarantee of equal rights meant that sit-inners and freedom riders were not criminals but rather upholders of constitutional law—even if southern sheriffs threw them in jail. For the opponents of apartheid, racism was a violation of internationally guaranteed human rights. For war resisters from Vietnam to Iraq, the national and international law forbidding war crimes defined their action not as interference with legal, democratic governments, but rather as a legal obligation of citizens. For the activists of Solidarity, the nonviolent revolution that overthrew Communism in Poland was not criminal sedition, but an effort to implement the international human and labor rights law ratified by their own government. As Jonathan Schell put it in the introduction to Adam Michnik's *Letters from Prison*, these agreements meant that the actions of Michnik and his associates were perfectly legal, "while the means used by the police and judiciary apparatus in Poland" were "in flagrant violation of international agreements."[7]

These examples seem paradoxical. On the one hand, the movement participants appear to be resisting the constituted law and the officials charged with implementing it. On the other, they are claiming to act on the basis of law, in fact to be implementing the law themselves against the opposition of lawless states.

Law professor and historian James Gray Pope has developed a concept of "constitutional insurgency" to understand such cases.[8] A constitutional insurgency is a social movement that rejects current constitutional doctrine, but that "rather than repudiating the Constitution altogether, draws on it for inspiration and justification." Such an insurgency "unabashedly confronts official legal institutions with an outsider perspective that is either absent from or marginalized in official constitutional discourse." On the basis of its own interpretation of the US Constitution, such an insurgency "goes outside the formally recognized channels of representative politics to exercise direct popular power, for example through extralegal assemblies, mass protests, strikes, and boycotts." It may hold such actions legal, even though the established courts condemn and punish them.

Pope recounts how the American labor movement long insisted that the right to strike was protected by the Thirteenth Amendment to the US Constitution, which forbids any form of "involuntary servitude." Injunctions to limit strikes were therefore unconstitutional. While courts disregarded this claim, the radical Industrial Workers of the World told its members to "disobey and treat with contempt all judicial injunctions," and the "normally staid" American Federation of Labor maintained that a worker confronted with an unconstitutional injunction had an imperative duty to "refuse obedience and to take whatever consequences may ensue."

Such insurgencies do not fit neatly into either the idea of a revolutionary overthrow of the government or of reforms conducted within the limits of legally permissible action as courts currently interpret them. In practice, social movements have long enacted a middle way between the constitutional discontinuity of revolution on the one hand and reform that fails to challenge the legitimacy of current legal structures on the other. The concept of constitutional insurgency explains how this can be.

The idea of a constitutional insurgency fits well with the practice of nonviolent direct action, which is extra-constitutional and yet not aimed at overthrowing the government per se. Indeed, when Gandhi said during the civil disobedience campaign that "sedition has become my religion," it might have been equally apt to say that he had become a constitutional insurgent, fighting for rights that English law enshrined but that its practice was denying to India. (As conservative historians are wont to point out, the American Revolution also began as a struggle for "the rights of Englishmen.")

Why a Global Insurgency?

The destruction of the climate by GHGs is produced in specific locations throughout the earth; it affects specific locations in every part of the globe; it can only be corrected through global solutions implemented in specific locations. The world order that perpetuates climate destruction is global, but it is produced and reproduced in specific locations around the world. The whole must be changed in order to change the parts; changing the parts is necessary to change the whole.

A global insurgency is not so much an effort to overthrow one or another government as it is an effort to transform the world order. That may seem like a tall order, but in some ways transforming the world order is easier than transforming the social and political order of individual nations. World orders are notoriously disorderly and fluid; their structure is maintained primarily by the mutual jostling of independent power centers. They change all the time: Where is the division of the world between two Cold War rivals or the global Keynesian economic regulation of fifty years ago? Moreover, unlike national governments operating under constitutions with officials chosen by elections, the world order has not the slightest claim to legitimacy. No electorate has ever consented to superpower rivalry or global neoliberalism—or destruction of the earth's climate.

It is against this illegitimate but mutable world order that a climate protection insurgency is ultimately aimed.

CHAPTER 7
CLIMATE PROTECTION AS A LEGAL DUTY

The climate protection movement has had no difficulty in articulating the moral dimensions of climate destruction, but it has had a harder time finding a legal frame to define its objectives and legitimate its actions. Existing environmental laws and treaties have proven inadequate to meet the challenge of climate change. Indeed, two-thirds of the GHG emissions in the United States are emitted pursuant to government-issued permits.[1] Recently, however, an ancient legal principle known in the United States as the public trust doctrine may be emerging to play a role. The application of public trust principles to climate protection is laid out in a series of recent lawsuits against national and state governments. Whether or not courts decide to enforce them, public trust principles can provide a powerful basis for a constitutional insurgency.

Governments have long served as trustees for rights held in common by the people. In American law, this role is defined by the public trust doctrine, under which the state serves as trustee on behalf of the present and future generations of its citizens. The principle is recognized today in both common law and civil law systems in countries ranging from South Africa and the Philippines to the United States and India.

The rationale for the public trust doctrine is that the unorganized public has sovereign ownership interests. Even if the state holds title, the public is the "beneficial owner." As trustee, the state has a "fiduciary duty" to the owner—a legal duty to act solely in the owners' interest with "the highest duty of care."[2] Government officials may not, consistent with the public trust, act for the primary purpose of benefitting a corporation—no matter how politically powerful it may be.

The principle underlying the public trust doctrine has roots and analogues in ancient societies in Europe, East Asia, and Africa, and also from Islamic to Native American cultures.[3] It was codified in the *Institutes of Justinian*, issued by the Roman Emperor in AD 535. The code defined the concept of "res communes" (common things). "By the law of nature these things are common to mankind—the air, running water, the sea and consequently the shores of the sea." The right of fishing in the sea from the shore "belongs to all men."[4] The Justinian code distinguished such "res communes" from "res publicae," things that belong to the state. The state cannot sell or destroy that which belongs not to it but to the people.

The public trust doctrine is a principle of national law in the United States and many other countries. International law, furthermore, recognizes geographical areas that lie outside of the political reach of any one nation-state—specifically, the high seas, the atmosphere, Antarctica, and outer space—as "global commons" governed by the principle that they are "the common heritage of humankind."[5] However, there has been no effective legal vehicle for asserting our right not to have our common environment destroyed.

Atmospheric Public Trust Litigation

On Mother's Day, 2011, the youth organization Kids vs. Global Warming organized the "iMatter March" of young people in 160 communities in forty-five countries, including the United States, Russia, Brazil, New Zealand, and Great Britain.[6] Concurrently, the Atmospheric Trust Litigation Project brought suits and petitions on behalf of young people in all fifty US states and the federal government to require them to fulfill their obligation to protect the atmosphere as a common property.[7] Speaking to one of the rallies, sixteen-year-old Alec Loorz, founder of Kids vs. Global Warming and lead plaintiff in the federal lawsuit, put the argument succinctly:

> Today, I and other fellow young people are suing the government, for handing over our future to unjust fossil fuel industries, and ignoring the right of our children to inherit the planet that has sustained all of civilization.
>
> The government has a legal responsibility to protect the future for our children. So we are demanding that they recognize the atmosphere as a commons that needs to be preserved, and commit to a plan to reduce emissions to a safe level.
>
> The plaintiffs and petitioners on all the cases are young people. We are standing up for our future.[8]

The suits argue that the atmosphere belongs in common to all people of current and future generations. Governments serve them as trustees but do not themselves own the atmosphere. "The State government may not manage the atmospheric trust resource in a manner that substantially impairs the public interest in a healthy atmosphere."[9] Governments have a sovereign duty to prevent substantial impairment of crucial public resources. The suits seek declarative judgment applying the public trust doctrine to the earth's atmosphere and ask the courts to issue injunctions ordering federal and state governments to reduce carbon emissions to fulfill their duty to protect it.[10] Similar suits are projected for countries around the world.

While so far the courts have turned down most of these atmospheric public trust suits, the decisions are being appealed. On October 3, 2013, the Supreme Court of Alaska became the first state supreme court to hear such an appeal.[11]

University of Oregon law professor Mary Christina Wood lays out the basis for such atmospheric trust litigation in an extensive legal article.[12] The public trust doctrine is

> a declaration of public property rights as originally and inherently reserved through the peoples' social contract with their sovereign governments. Under this principle, the public holds a perpetual common property interest in crucial natural resources. Government, as trustee, must act in a fiduciary capacity to protect such natural assets for the beneficiaries of the trust, which include both present and future generations of citizens.[13]

The trustee has "an active duty of vigilance to 'prevent decay or waste' to the asset." "Waste" means permanent damage. If the asset is wasted in the interest of one generation of beneficiaries over future generations, it is in effect an act of "generational theft."[14]

The basic principles of public trust law were set out in the 1892 US Supreme Court decision *Illinois Central Railroad Co. v. Illinois*. The Illinois legislature had conveyed its title to the Chicago shoreline of Lake Michigan to the Illinois Central Railroad, one of the richest and most powerful corporations of its day. The Court ruled that the legislature had no power to do so, because the state was not the owner of the property, but rather the trustee for the citizens, the present and future generations of the public. "Such property is held by the state, by virtue of its sovereignty, in trust for the public." The reason is that the ownership of the navigable waters of the harbor and the lands under them is "a subject of public concern to the whole people of the state." Therefore, "the trust with which they are held" is "governmental, and cannot be alienated."

The public trust doctrine limits the power of government based on the fundamental principles of constitutional democracy. As the US Supreme Court wrote in 1896 in another formative case, *Geer v. Connecticut*, "The power or control lodged in the State, resulting from this common ownership, is to be exercised, like all other powers of government, as a trust for the benefit of the people, and not as a prerogative for the advantage of the government as distinct from the people, or for the benefit of private individuals as distinguished from the public good." As Wood explains, "Government trustees may not allocate rights to destroy what the people rightly own for themselves and their posterity."

Courts are now applying the public trust principle to questions that bear directly on climate protection. The Pennsylvania Supreme Court in December 2013 overturned a law that prevented local communities from blocking fracking for natural gas. The plurality opinion held that public natural resources are owned in common by the people, including future generations. Because the state is the trustee of these resources, it has a fiduciary duty to "conserve and maintain" them. The state has "a duty to refrain from permitting or encouraging the degradation, diminution, or depletion of public natural resources."[15]

When a trust asset crosses the boundaries of sovereign governments, all sovereigns with jurisdiction over the natural territory of the asset have legitimate property claims to the resource. So all nations on earth are "co-tenant trustees" of the global atmosphere. They have a duty not to commit waste to the common property.[16]

Two legal duties arise from this relationship. First is "the sovereign duty that each government, as trustee, has towards its own citizens to protect the atmospheric asset and prohibit waste of their natural inheritance." Second is "the duty owed by each nation towards all other nations, arising from the sovereign co-tenancy relationship, to prevent waste to their common asset, the atmosphere."[17]

Violation of these duties can lead to two legal claims. "Citizen beneficiaries" can bring actions against their governmental trustees for "failing to protect their natural trust." Additionally, one sovereign trustee can bring actions against others "for committing waste to common property."[18]

Fair Remedies

If a court upheld such claims against the co-tenant trustees—the nations of the world—what could it order them to do? The prime questions on which international negotiations for climate protection have faltered are how much GHG emissions should be cut, and how quickly; and how the burden of protection should be distributed. Mary Christina Wood addresses these questions on the basis of the principles that courts normally apply to public trust obligations.

According to leading climate scientists such as Dr. James Hansen, reducing atmospheric carbon to 350 ppm or less is necessary to avoid catastrophic climate change. Taking 2012 as a baseline, an annual global decline of 6 percent in fossil fuel emissions, combined with the extraction of 100 gigatons of carbon dioxide through reforestation and improved forestry and agriculture, would lower the atmospheric concentration of carbon dioxide to 350 ppm by the end of the century.[19] This goal probably will require reaching near-zero carbon emissions by around 2050. Therefore, courts must impose a timeline with an endpoint of near-zero emissions.

While these scientific calculations indicate what the world as a whole must do, different countries vary widely both in their contribution to wasting the atmosphere and in their ability to pay for climate protection measures. When waste of a common asset occurs, courts apportion "fair shares" of the costs of remediation to the various responsible parties. Mary Christina Wood identifies five factors that courts would need to weigh in assigning to countries their fair shares to remedy global warming:[20]

- Global share of current carbon emissions.
- Historical share of emissions.
- Per capita emissions. Each American uses nearly 20 metric tons of carbon dioxide emissions on average, compared to 1.16 metric tons for each citizen of India.
- Purpose of the emissions. Priority should be given to meeting basic human needs, then to creating new infrastructure for a low-carbon society, and least to providing nonessential and frivolous luxuries.
- Recalcitrance of the sovereign in taking responsibility for the country's carbon pollution.

A widely cited study called the *Greenhouse Development Rights Framework* (GDRF), prepared by the Stockholm Environmental Institute and EcoEquity, has already quantified the first four of these factors. The study evaluates the responsibility and capacity of every country for GHG reduction. Responsibility is measured by the country's cumulative GHG emissions since 1990, while capacity is based on the ability of a country to reduce emissions without threatening the basic survival of its people. Capacity is derived from the national income, but it doesn't count income required for the necessities of daily life. The measure thus takes into account the unequal distribution of income within as well as between countries, providing that the very poor aren't required to pay for a problem they have done little to create. The bottom line is an evaluation of the fair share of GHG reduction for each country. Taken together, the shares add up to the cuts scientists estimate are necessary to reach 350 ppm by the end of the twenty-first century.[21]

If courts find the co-tenant trustees wasting the atmospheric public trust in violation of their fiduciary duty to protect it, what remedies should they offer? First, according to Wood, courts should issue declaratory judgments expressing the fiduciary obligation of all governments to protect the atmosphere as a commonly shared asset, to be realized in a scientific prescription for carbon reduction to levels below 350 ppm.

Second, courts should issue injunctions requiring all agencies of government to take the measures necessary to realize this duty. Those injunctions may require "carbon accountings," which quantify carbon emissions and track their reduction over time. They may include "enforceable carbon budgets" that set quantifiable mileposts and may also require periodic progress reports.

A court need not tell the government how to realize its duty, but it can require a government to present a plan demonstrating how it will do so.[22] If the plan is not carried out, a court can itself issue injunctions prohibiting specific wasting activities, such as issuing of permits for new coal-fired power plants or setting excessive air pollution quotas. Ultimately, it can find disobedient government officials in contempt of court.

Waste of the atmosphere is largely conducted by private businesses. In trust law, "trustees have the affirmative duty to recoup monetary damages against third parties that destroy trust assets." In a public trust, "all sovereigns theoretically have grounds for recovering damages from third parties who destroy the trust." Requiring fossil fuel companies to pay damages for the colossal waste they have committed on the public trust would go a long way toward paying for the transition to a low-carbon economy.[23]

An Insurgency to Protect the Atmospheric Public Trust?

As compelling as the logic of the "atmospheric public trust" argument may be, it is easy to imagine that many American courts will refuse to force governments to meet such obligations. In a brief to dismiss the Kansas suit, lawyers called the claim "a child's wish for a better world," which is not something a court can do much about. "No order issued by the District Court of Shawnee County can hold back global warming, any more than King Canute could order the tide to recede." "It's Hail Mary pass litigation," according to Michael Gerrard, director of the Center for Climate Change Law at Columbia University's law school.[24] The sad fact is that virtually all the governments on earth—and their legal systems—are deeply corrupted by the very forces that gain from wasting the public trust. They exercise illegitimate power without regard to their obligations to those they claim to represent, let alone to the trust beneficiaries of other lands and future generations to whom they also owe "the highest duty of care."

Indeed, the effort to halt global warming by suing in governmental courts to enforce the public trust doctrine would seem to run up against the entire world order and confront all the obstacles that have blocked climate protection until now.

However, protecting the atmosphere is not just a matter for governments. The failure of governments to protect the global commons is currently leading the climate protection movement to turn to mass civil disobedience, as witnessed by the campaigns against the Keystone XL pipeline, mountaintop-removal coal mining, coal-fired power plants, natural gas fracking, and Arctic oil extraction. Viewed from the perspective of the public trust doctrine, these actions are far from lawless. Indeed, they embody the effort of people around the world to assert their right and responsibility to protect the global commons. These actions show people acting in an emergency situation on an evident necessity and represent people stepping in to provide law enforcement where corrupt and illegitimate governments have failed to meet their responsibility to do so. They are confronting lawlessness by taking the law into their own hands.

In 2007, five activists climbed the chimney of the Kingsnorth coal-fired power plant in Kent, England, painted a protest message on it, and tried to shut down the plant. The government charged them and an associate with causing more than $50,000 worth of damage. The protesters admitted that they had entered and tried to shut down the plant, but argued that they were legally justified in doing so because they were trying to prevent climate change from causing far greater damage to property around the world.

In an eight-day trial, NASA's James Hansen told the court that humanity was in "grave peril" and that "somebody needs to step forward and say there has to be a moratorium" against coal-fired plants. Hansen testified that the carbon dioxide emitted from the plant could be responsible for the extinction of up to four hundred species. Other witnesses during the trial described property that was in peril of destruction from climate change. The Pacific island state of Tuvalu and parts of Greenland were at risk from rising sea level—as were areas right there in Kent, England.[25]

In his summation at the end of the trial, the judge said that the case centered on whether or not the protesters had a "lawful excuse" for their actions. He told the jury of nine men and three women that to use a lawful excuse defense the protestors had to prove that their action was due to an immediate need to protect property belonging to another. By majority vote the jury voted to clear the Kingsnorth Six.[26]

In October 2014, Alec Johnson, aka "Climate Hawk," was tried for locking himself to an excavator on a Keystone XL pipeline route in Atoka, Oklahoma. If convicted, he could have faced up to two years in the Atoka County Jail. He may be the first climate protester to use the public trust doctrine as the basis for a necessity defense.

In a speech on the day of the 2014 Peoples Climate March, Johnson said, "When it comes to our commons, to our public property, we the people have rights in a Public Trust." Public Trust doctrine "assures us that we have rights when it comes to how our public commons are administered by any trustees placed in charge of it." We the people are "armed" by such legal doctrine. We now "demand our environmental institutions and agencies recognize their responsibilities as trustees and exercise their fiduciary responsibility to act with 'the highest duty of care,' to ensure the sustained resource abundance necessary for society's endurance."

The pipeline itself is "proof that our trustees have failed in their fiduciary duty to 'We the People.'" The presence of any of this pipeline in the ground "incontestably proves that the states of Oklahoma and Texas have failed." And the US government has failed to discharge "the highest duty of care," on behalf of this and future generations.[27]

In a web appeal for support, Alec Johnson wrote, "I won't plead guilty for taking direct action against extreme energy madness. Enforcing our children's rights to climate justice is no crime." The jury convicted without incarceration; Johnson commented, it "felt like a victory."[28]

Ultimately, protecting the public trust is a duty we all as members of the public owe ourselves, each other, and future generations. As the Supreme Court of India put it, "Today, every person exercising his or her right to use the air, water, or land and associated natural ecosystems has the obligation to secure for the rest of us the right to live or otherwise use that same resource or property for the long term and enjoyment by future generations."[29] Mary Christina Wood writes that the fiduciary obligation of all governments, as trustees, to protect the atmosphere as a commonly shared asset is "enforceable by the citizen beneficiaries of the trust representing present and future generations."[30]

If the courts fail to provide such protection, have not "citizen beneficiaries" a right and duty to enforce that obligation by other means?[31] In his *Second Treatise of Civil Government*, perhaps the single greatest influence on the shaping of American government, John Locke wrote that "whenever the legislators endeavour to take away, and destroy the property of the people" they "put themselves into a state of war with the people," who are thereupon "absolved from any farther obedience" and are "left to the common refuge, which God hath provided for all men, against force and violence." When legislators "either by ambition, fear, folly or corruption, endeavour to grasp themselves, or put into the hands of any other, an absolute power over the lives, liberties, and estates of the people" by their "breach of trust" they "forfeit the power the people had put into their hands for quite contrary ends." Then that power "devolves to the people," who have a right to "resume their original liberty" and, "by the establishment of a new legislative [power], (such as they shall think fit) provide for their own safety and security."

CHAPTER 8
MAKING A COUNTRY CLIMATE-SAFE

The climate protection movement has had little trouble portraying the dangers of climate change, but it has had far more difficulty providing a credible plan to transition to a climate-safe economy without mass unemployment and economic catastrophe. One thoughtful and well-known climate activist, asked what would happen if protestors managed to shut a coal plant down permanently, replied, "If the question is, 'What do we do after we shut it down tomorrow,' somebody else will have to figure that out."[1] Facile assurances that climate protection will produce more jobs than it destroys have not filled the need for a concrete pathway that protects people as well as the climate.

This reticence is due in part to mind-sets that obscure a realistic alternative vision. Neither the market, nor local initiatives, nor consumer restraint, nor revolution can alone provide a credible path to climate protection. Rapid reduction of GHG emissions cannot be achieved without breaking out of the neoliberal shibboleths that have dominated public policy for the past thirty years. Nor can it be realized simply by local initiatives to create resilient communities or by persuading individuals to make do with less. Revolutions producing new climate-protecting regimes in each of the world's countries seem unlikely in the time frame necessary to forestall devastating climate change. Are other, more feasible visions conceivable?

Suppose a compelling force—either legal or popular—required governments to fulfill their duty to the atmospheric public trust. If, due to some combination of political decision making, legal compulsion, public demand, international pressure, and insurgent challenge, a country decided to reduce its

GHG emissions to a level compatible with reaching 350 ppm globally, how could it do so? Let's look at the United States as an example. What kind of climate action plan could realize its public trust duties?

As we have seen, to reach 350 ppm by the end of the century—starting from 2012 as a baseline—will require a global reduction of 6 percent per year in fossil fuel emissions, combined with the extraction of 100 gigatons of carbon dioxide from the atmosphere.[2] Global carbon emissions will need to be near zero by around 2050. The fair share of reduction would be substantially higher for wealthy countries like the United States that have contributed large amounts of GHGs in the past.

Studies show that such a reduction is technically feasible and suggest various pathways to achieve it.[3] While it can be achieved based on commercial technologies available now, rapid expansion of research and markets will likely lead to very rapid improvement in technology along the way. Reduction can be based on conversion to low-carbon energy technologies, more efficient use of energy, and reduced energy demand. It will not require nuclear energy, geo-engineering, or carbon capture and storage, each of which is likely to be far slower, more costly, and environmentally dangerous than low-carbon energy, energy efficiency, and conservation. Because rapidly reaching 350 ppm requires rapid conversion to renewables and reduced demand, there is only a small need for natural gas as a transitional fuel.

The most important targets for GHG reduction are electricity, transportation, and buildings. Electricity produced by fossil fuels, the largest single source of GHGs, can be replaced by renewable energy, energy efficiency, conservation, improved transmission lines, and new energy-storage technologies. Petroleum-based private transportation can be replaced with public transport and cars fueled with renewable energy and biofuels. Freight transportation can be converted to rail transport and electric and biofuel vehicles. The carbon footprint of nearly all buildings can be sharply reduced through insulation, weatherization, cogeneration, and solar and geothermal heating, cooling, and hot water. Many other strategies, ranging from industrial redesign to integrating urban and transportation planning and from expanding forests to reducing fossil fuel use in farming, will also contribute.

There are three main approaches to GHG reduction. The first, which has dominated climate legislation and treaty negotiation, consists of "putting a price on carbon" to discourage GHG emissions through taxation, fees, cap-and-trade systems with markets for emission quotas, or similar means. The second route, which is widely discussed and frequently implemented on a small scale, consists of local, often community-based, initiatives designed to produce renewable energy and reduce energy consumption on a decentralized basis. The third approach, perhaps less often delineated by proponents than excoriated by opponents, consists of a government-led, centralized program based on economic planning, public investment, resource mobilization, and direct government intervention in economic decisions. While these are often presented as alternative choices, rapid reduction of GHG emissions will undoubtedly require all three.

Mobilization—The World War II Model

The government-led approach often uses the economic mobilization for World War II as a touchstone—either to show the feasibility of rapid and massive economic change, or to reveal the evils of a "command economy" that interferes radically with the private market. While Al Gore and others have cited World War II mobilization as a possible model for climate protection, there have been few in-depth presentations of how such an approach might work in practice. Fortunately, two recent papers by Laurence L. Delina and Mark Diesendorf examine the World War II mobilization and suggest what lessons—positive and negative—can be drawn from it for rapid reduction of GHG emissions.[4] They argue that climate protection may well require government-led mobilization on the scope and scale of World War II to solve many similar problems, but that the particular form of such mobilization will need to be different both because of the differences in purpose and because the projects pose rather different problems.

The scale and scope of US economic mobilization for World War II was truly impressive.[5] US military spending rose from less than $2 billion in 1940 to more than $90 billion in 1944—an increase of more than $1 trillion in 2010 dollars. In the five years of the war, the United States produced 300,000 planes, 100,000 ships, and 20 million rifles. During wartime, investment in research and development produced radically new technologies; the United States spent more than $20 billion in 2008 dollars and directly and indirectly employed more than 100,000 people on the Manhattan Project alone to produce the first atomic bomb.

War production was based on strategies for finance, labor, and governance.

Finance: The huge and rapidly growing US military expenditure was paid for primarily by taxes and borrowing. US government tax collections grew from less than $9 billion in 1941 to $45 billion in 1945.[6] For their part, 85 million Americans bought $185 billion in war bonds and similar securities—more than $2 trillion in 2010 dollars.

Labor: The number of Americans employed outside the military rose by 7.7 million between 1939 and 1944, even while millions more left the civilian labor force for the military. Government boards redirected workers to military production, sometimes by threatening to draft them otherwise. Women entered the industrial workforce on an unprecedented scale and government provided training for millions of workers. The National War Labor Board set wages and required employers to bargain collectively with their employees' unions. Government built housing and provided healthcare and childcare for war workers.

Governance: The US government established the War Production Board, chaired by a "war czar," which, along with more than 160 other war agencies, took over direction of much of the US economy, functionally replacing much of the private market. The US government directly controlled more than 40 percent of the country's production of goods and services. The government set production goals, supervised and managed industry, and determined which producers could get essential materials. It operated important industries and even paid for and owned war plants. The government could force companies to agree to government contracts and requisition private property; it could also halt production that interfered with military needs: From 1942 to 1944, the government simply halted production of private cars. Washington established financial and banking controls and regulated the economy through fiscal and monetary policy, wage and price controls, and rationing. It provided highly profitable contracts, subsidies, and tax rebates to private companies, but it also imposed a tax on excess profits.

Mobilization for Climate Protection

The scale and scope of change necessary to reach the climate goal of 350 ppm is surely comparable to that of mobilization for World War II. It will involve a great deal of new production, and some current production will need to be halted. The nature of the task is rather different, however, as the purpose is not just to ramp up the quantity of production, or simply to shift it to a new set of products. While both are necessary, the task goes far beyond that to a qualitative transformation to an economy—and society—based on very different technologies. The task will take far longer, will require longer-term planning, and must be accomplished in a way that is permanently sustainable. However, like war mobilization, it will require strategies for finance, labor, and governance.[7]

Finance: The starting context of climate protection mobilization is the massive failure of private markets to invest in renewable energy, energy efficiency, and energy-demand reduction, which will require major public investment to correct.[8] Such mobilization will also require large-scale, long-term planned development of new infrastructure and other systems far beyond the capacity of private corporations. Over time the cost of economic transformation will fall, both because low-GHG energy capacity is expensive to construct but cheap to run, and also because its costs will inevitably fall due to economies of scale of mass production and improved production technologies. The initial costs of transformation, however, will be high.

Today, as at the outset of World War II, the US economy is mired in the aftermath of a severe economic decline with vast quantities of underutilized resources. By very conservative estimates, the economy is now operating at only 95 percent of its potential. According to the Federal Reserve, the capacity utilization rate for total industry is still under 80 percent.[9] Were the United States at full employment, the economy would be producing $800+ billion a year more than it currently is, generating the resources we need to convert to renewable energy and provide a just transition for workers, communities, and carbon-dependent regions. A public investment–led recovery would stimulate private investment by providing a secure market for the products of such investment. Currently there is more than $1 trillion of cash on corporate balance sheets, with $1 trillion in excess bank reserves parked at the Federal Reserve.

Public borrowing through bond sales can provide substantial and inexpensive funds due to the currently low borrowing rates for government debt. If need be, the Fed can simply buy infrastructure bonds, just as it did with Treasury securities in 1940 to finance the war. Public-purpose banks, credit unions, and investment and loan funds can provide more decentralized financial resources, especially for smaller-scale and community-based projects.

Within a context of growing productive capacity, tax policy can help discourage carbon emissions while reversing our growing income inequality. Taxation of carbon emissions or "cap-and-dividend" programs can provide market incentives for conversion to lower GHG emissions. As a matter of justice—as well as to win the broadest popular political support—most if not all of the resulting revenue should be returned to the workers and consumers to compensate for higher gas prices and higher energy bills and to restore a more just distribution of income. Progressive taxation, particularly on carbon-wasting luxury goods like private jets, can counteract any negative effects on income equality. Such devices as energy-pricing incentives, user fees, and on-bill financing (which allows energy consumers to pay for energy-saving investments out of the resulting savings on their energy bills) can also play a role.

Another source for funding a transition to climate safety could be legal damages collected from corporations for the environmental damage they have committed. Governments may take legal action to recover "Natural Resource Damages"—as seen in the settlements for the Exxon Valdez and BP oil spills, for example. The Comprehensive Environmental Response, Compensation, and Liability Act of 1980 (CERCLA—known as the Superfund law) provides broad federal authority to clean up hazardous substance releases and authorizes the US Environmental Protection Agency (EPA) to compel the parties responsible to pay for the cleanup—even if the releases happened long before the legislation was passed. Comparable legislation could hold major fossil fuel producers and emitters responsible for their colossal damage to the atmosphere—and the colossal cost of remediating it.[10]

Labor: Nearly 12 million Americans are officially unemployed today; more than 8 million want full-time work but are only employed part-time; 2.6 million people want to work and have sought work within the past year but are not currently looking for work.[11] So a labor reserve of more than 20 million workers is available to go to work protecting the climate. However, ways will be needed to redirect workers to the growing employment sectors. During World War II, this was done by the War Labor Board, which actively recruited workers to regions and industries where they were most needed; the board also controlled wages to limit competitive bidding for scarce labor. Government took the leading role in the rapid expansion of education and training for the new workforce.[12]

New labor policies will be needed both to protect the relatively small number of workers who will lose jobs in fossil fuel–related industries and to ensure popular support for the transformation by providing improving conditions of life for the population. As in World War II, for the process to be generally accepted as fair will require an incomes policy that boosts lower income extremes and restrains higher ones. A Nordic-style welfare-state system—one based on full employment, combining a high level of income for the unemployed with strong support for retraining and new jobs—will be necessary to allay fears that change will lead to disaster for workers.[13] Public planning, investment, and incentives for new employment opportunities in affected regions, industries, and occupations can play a similar role. As in World War II, the right of workers to organize and bargain collectively with their employers will be essential to ensure popular participation in the mobilization and protect workers from abuse.

Governance: Government action will be necessary to implement many of these transformations. Delina and Diesendorf list such actions as establishing financial incentives and disincentives; raising capital; implementing labor strategies; organizing funding for infrastructure such as transmission lines, railways, and pipelines; funding R&D; setting and monitoring energy-efficiency standards for buildings, appliances, and equipment; training and retraining professionals and tradespeople; and setting industrial location policies. Further, the multifaceted activities of federal agencies, state and municipal governments, corporations, and civil society groups will need to be coordinated to capture synergisms and prevent them from undermining each other.

Such coordination, as in World War II, will require a central governmental authority. However, because of the extended period of transition, measures are necessary to prevent such an authority from deviating from its intended purpose either for its own aggrandizement or that of other social forces. The United States doesn't need another Pentagon or NSA provided with vast powers and resources but no genuine accountability.

Delina and Diesendorf propose two agencies, independent of each other, to lead the transition to a low-GHG economy. The first, following the general model of the War Production Board, would have overall responsibility for GHG reduction. It would "conduct technical requirement studies, set and enforce production goals for renewable energy technologies, institute efficient contracting procedures, cut through inertia and 'red tape' inhibiting institutional changes, and serve as the coordinating agency for all transition activities."

Legislation would also establish a separate, countervailing institution to play a planning and watchdog role. This agency would be independent of the executive branch and above the transition agency; it would report to Congress and the public. While Delina and Diesendorf don't spell out its powers and procedures, they would presumably include defining GHG-reduction targets and timetables; laying out a national climate action plan; ensuring transparency in the actions of the climate-mobilization authority; identifying problems and failures; and initiating needed course corrections. It would set time limits on executive authority, provide checks and balances, scrutinize government actions, and "ensure that the government/executive sticks to its transition mandate."

Although government will have a leading role, markets will have a crucial role as well. Most economic activities will continue to be coordinated through markets, albeit ones affected by new public policies. Market-based approaches, such as energy price incentives, carbon taxes, fees, and/or quotas, will help redirect production and investment to low-GHG technologies and products in the myriad areas not covered by direct government policies.

Finally, civil society organizations will have at least as critical a role. Today, a large swath of community-based, local, and regional programs initiated from below are already engaged in promoting the transition to a climate-safe economy and society. Even in a government-led transition, they can on their own initiative implement community-based renewables, energy-use reduction, mobilization of funding, and new patterns of consumption. Perhaps most important, they can provide both popular support for transition and a means to hold the institutions of transition accountable.

Many such climate-protecting activities are already under way, albeit in unconcerted form. The US government reorganized the auto industry in a way that produced cars with sharply reduced carbon emissions. Public mobilization, combined with EPA regulation and economic forces, has virtually ended the building of new coal-fired power plants and led to the closing of more than 140 existing ones.[14] In Germany, energy-pricing policies have led to massive expansion of renewables—25 percent of Germany's electricity now comes from solar, wind, and biomass.[15] Moreover, decentralized civil society initiatives are weatherizing houses, installing solar collectors, and pressuring governments and businesses at every level to transition to a low-carbon basis. These activities provide a seedbed from which more extensive climate protection measures can grow. They also initiate a learning curve that will continue—facilitating continuing course correction—until a safe GHG level is reached.

How does this approach relate to the longstanding debate between the advocates of economic growth and those who believe human economic activity needs to be stabilized or reduced in order to reduce its impact on the earth? It is different from either in that it is neither pro- nor anti-growth per se. Such an approach presumes that there are some things that we should grow, like human health and the security of the environment; there are other things, conversely, that we should shrink, such as dangerous labor and GHG emissions. The goal is to fit our economic activity to the real needs of people and the environment, which requires both regulating our overall level of economic activity and radically shifting what we produce and how we produce it. Combining full employment and climate protection represents not growth for the sake of growth, but increasing what benefits us and shrinking that which is doing us harm.[16]

CHAPTER 9

A GLOBAL TRUST FUND
FOR THE GLOBAL PUBLIC TRUST

The issue on which international climate negotiations have visibly foundered is the distribution of the costs of climate protection between developed and developing countries. This includes "development space"—whether and how much developing countries should have to restrict their GHG emissions, given the "free ride" developed countries have been given to emit greenhouse gases without restriction for the past two centuries. It also includes the closely connected issue of whether and how much developed countries should contribute toward GHG reduction in developing countries.

These discussions take place in the context of an ongoing crisis in the global economy. Neoliberal doctrine has called for a global austerity that generates massive unemployment of human and material resources and causes sovereign debt crises and a race to the bottom in working, living, and environmental conditions in countries around the world. The paradox of this economic squeeze is that the world's human and material resources are being placed "out of service" at the very time they are desperately needed to fight global warming. The current world order blocks the mobilization for climate protection of resources that currently lie dormant. A central goal of a constitutional insurgency should be to correct the wasting of the atmospheric public trust by spurring the mobilization of the underutilized global human and material resources needed to reduce atmospheric GHGs below 350 ppm.

Many of the flaws of the current world order have converged around the question of paying for poorer countries' climate

101

protection. The absence of strong institutions representing global interests makes it nearly impossible to come to a workable consensus on climate policy, even though it would be in the interest of all parties to do so. The sovereignty of individual nations makes it impossible to impose global taxes to meet common global needs. Competition among states means that what is interpreted as a gain for one is experienced as a loss for others. The hegemony of the richest and most powerful, and the drive of others to challenge it, reduces questions of global justice to a power struggle among coalitions of interests. The dominant neoliberal ideology opposes any effort to invest for the long-term public interest at the expense of short-term private profit. A global economy with no means to regulate its level of economic activity makes expenditures for global climate protection appear to be a drain on individual and national well-being. Meanwhile, vast resources are squandered by military spending and war to aggrandize the wealth and power of sovereign states and coalitions.

The "atmospheric public trust" approach laid out previously in Chapter 7, "Climate Protection as a Legal Duty"—whether imposed by courts, democratic public pressure, or nonviolent insurgency—includes a way of allocating global rights and responsibilities for climate protection based on the legal principles that govern public trusts. This allocation has been spelled out in the Global Development Rights Framework, which evaluates countries on the basis of both their historical responsibility for the problem and their capability to help solve it. The GRDF goes beyond allocation between rich as opposed to poor countries by considering the distribution of income within countries, factoring in the responsibility and capability of rich and poor individuals within each country.

International negotiations have centered on the mechanism of a global fund to pay for the costs to poor countries of investing in climate protection and compensating for the benefits they forego by not polluting. Negotiations on such a fund have broken down over the questions of how big it should be, how the money should be used, how it should be paid for, who should control it, and how it would work.

How big should such a trust fund be? At the least it should be big enough to mobilize all the unused human and material resources that can be applied to climate protection. According to the International Labor Organization, 200 million workers were unemployed worldwide in 2013 and the number continues to rise years after the start of the so-called economic recovery. Unspent cash in the accounts of large enterprises had reached $5 trillion.[1] While estimates vary, a rough idea of the scale needed for rapid climate-change mitigation is given by a 2013 study sponsored by the World Economic Forum (WEF): This study concluded that at least $0.7 trillion needs to be invested annually beyond current levels "to limit the global average temperature increase to 2°C above pre-industrial levels."[2] So let us assume that at the least $0.7 trillion—somewhere between 1 and 2 percent of global GDP—can be effectively invested yearly in climate protection worldwide.

How should the money be used? The $0.7 trillion proposed by the WEF study includes investment needed for "clean-energy infrastructure, sustainable and low-carbon transport, energy efficiency in buildings and industry, and for forestry." Investment for climate change adaptation is not included.[3]

Where can the money come from? As with funding for national programs, global funding can come from taxing, borrowing, recovery of damages, and a global equivalent to fiscal policy. Here again the primary barriers have been the unwillingness of national governments to spend their money for a common global purpose, even one that is essential to their and their populations' survival, and the neoliberal austerity policies that demand social resources be left to languish if they cannot be used for debt repayment or private profit.

A tax on carbon emissions would provide an obvious source of funds; it would provide an incentive to reduce GHGs at the same time that it helped finance the effort to do so. Also widely advocated is a financial transactions tax, also known as the Robin Hood Tax, which would impose a small charge, perhaps one-half of 1 percent, on all financial transactions. The tax could be instituted by individual countries, by international agreement, or by an international institution. Such a tax has been widely advocated not only for the funds it could raise for climate protection and other public purposes, but also as a means to reduce the economic bubbles and busts generated by unrestrained financial speculation and to provide a basis for regulating the "shadow banking system."

The sale of climate bonds, perhaps guaranteed by a consortium of national governments, represents another source of funds. Investors could include ordinary citizens, governments, corporations, and other organizations. As activists demand disinvestment from fossil fuels, they can simultaneously demand that the liberated money be reinvested in such climate protection funds.

As we saw in the previous chapter, governments can seek damages from corporations for the waste their GHG pollution has inflicted on the atmospheric trust. Such damages could be a significant source of revenue for a global climate protection fund.

Paper Gold

Another source of funds could be a little-known international financial vehicle known as Special Drawing Rights (SDRs), often referred to as "paper gold."[4] In 1969, after a string of liquidity crises, the world's major governments agreed to create SDRs to increase global liquidity. Former World Bank chief economist Joseph Stiglitz explains that SDRs are "a kind of global money," which "countries agree to accept and exchange for dollars or other hard currencies." If countries acquire SDRs to add to the gold and foreign currency in their national reserves, those reserve funds can be put to use for other purposes instead of sitting idle.

Stiglitz proposed that SDRs or a new "global greenback" along similar lines be used to supplement other reserve currencies. They would be issued for investment in developing countries and for "global public goods" like environmental projects, health initiatives, and humanitarian assistance. They would have the added benefit of checking global deflation and would help countries with trade deficits avoid ruinous devaluations and runs on their currencies.

Until the Great Recession, such proposals received little public attention—indeed, few except international economists even knew SDRs existed. However, for a brief period beginning in 2009, world economic leaders advocated massive stimulus to the global economy. The IMF called for a global stimulus of 2 percent of the world's total product to sustain global demand in the economic downturn—about $1.2 trillion.[5] In this context, discussion of paper gold exploded. George Soros called for "trillions of dollars" in SDRs to be issued to fight the recession. British prime minister Gordon Brown campaigned for a new allocation of SDRs and the United States seemed to be warming to the idea.[6] SDRs could provide a global economic stimulus in much the same way that increasing a nation's money supply does nationally.[7]

As paying for developing countries' climate protection costs became more obviously central to a climate agreement, the idea of using SDRs to finance climate protection came to the fore. While the 2009 climate summit was foundering on the question of who would pay for developing countries' GHG restrictions, George Soros arrived in Copenhagen and proposed issuing US$100 billion in SDRs for a special green fund. The idea was soon advocated on behalf of the G-77 organization of developing countries in a passionate speech by Lumumba Di-Aping of

Sudan, the negotiator for the G-77. After the failure of Copenhagen, the IMF itself briefly toyed with the idea of using SDRs for a "Green Fund."[8] The idea went nowhere, however, as global economic leaders shifted their goal from recovery to austerity. Surely it is an idea whose time should come again.

Who should control a climate protection trust fund and how should it be administered? Such a fund should be controlled by a body specifically devoted to climate protection. The UNEP's authoritative scientific committee, the Intergovernmental Panel on Climate Change, could play a major role in setting criteria and evaluating the results.

How might it work? Countries would apply to the trust fund for SDRs and other financing that can be used solely to implement their national plans to reduce greenhouse gas emissions. The funds would be allocated based both on what help countries need to pay for their own climate protection costs and the importance of their efforts to global climate protection targets. In order to qualify, each country would be required to meet its obligation to reduce greenhouse gases, making the trust fund resources an incentive for countries to meet their public trust duties. Complete transparency in allocating and contracting can be a further condition for receiving trust fund resources. Funds could also be allocated, as Stiglitz has suggested, by competition among countries for the most worthwhile projects.

In addition to funding climate protection, such a fund in effect provides a vehicle for a countercyclical global macroeconomic policy. It provides needed stimulus for the global economy; promotes a carbon-reducing rather than carbon-expanding form of economic growth; mobilizes underutilized human and material resources; and partially solves the festering sovereign debt crisis and the economic, social, and political disorder it provokes. Without some such global macroeconomic regulation, national and corporate "beggar-your-neighbor"-type policies will drive countries to trade and military wars, perpetuating mass unemployment and the race to the bottom. A global climate protection fund can provide the growth point for broader global economic cooperation and regulation. And it can provide a framework for meeting human needs and wants and therefore maintaining popular support for climate protection.

To achieve its goals, a climate-protecting insurgency will have to transform the world order to make such economic cooperation possible.

CHAPTER 10
MOVEMENT ENFORCEMENT OF PUBLIC TRUST DUTIES

How can the transformation of the world order necessary to protect the climate be brought about? A plausible means, we have argued, is a global nonviolent insurgency enforcing the duty of governments to protect the atmospheric public trust. How can such an insurgency be created and how can it realize its goals?

Today's climate protection movement provides a starting point. With its global organization and civil disobedience challenge to established authorities, today's movement may already be on the way to becoming a global nonviolent insurgency. Moreover, while it has not yet declared its actions to be law enforcement protecting the public trust, that is in fact what they are doing.

Charting the future of a climate insurgency must be the ongoing work of many hands and many brains. This chapter aims not to lay out a comprehensive action program, but rather to provide a sketch of how an emerging global insurgency might go about protecting the atmospheric public trust.

Defining Climate Protection as Public Trust Protection

A first step to a constitutional insurgency is to define climate action as protection of the public trust. That can be done by any individual or group engaged in civil disobedience and other climate-protecting activities. They can argue that they are actually acting to enforce fundamental legal principles that protect the atmosphere as a public trust against the far greater harm being perpetrated by those who are committing waste against it. They can argue that they are protecting a common property right that they share with present and future generations. They can make a "necessity defense" that their action—say, blocking coal shipments—is necessary to prevent a far greater offense with far more serious consequences. They can make this case to judges, juries, and in what they say and write about their actions to the public. While no one should expect that courts will readily accept such an argument, it can be an effective way to redefine what climate action is all about.[1]

More broadly, the climate protection movement can incorporate the public trust argument as a central part of its message and its campaigns. When it attacks the Keystone XL pipeline or demands conversion to renewable energy, it can justify its action in part by the authorities' dereliction of their duty to protect the public trust and the ultimate right and duty of the people to prevent the wasting of their common property.

As the leaders of more than one hundred of the world's governments addressed the UN Climate Summit on September 23, 2014, people's organizations from around the world convened a Climate Justice Tribunal across the street from the UN to "indict political leaders and corporate polluters for their failure to protect our health, communities, and planet." Those testifying were "those living with the real and immediate impacts of climate change" and "people living on the frontlines of extractive industries that are contributing to climate change." Two of the witnesses were youth plaintiffs in the Our Children's Trust suits to legally compel states and nations to protect the climate based on the public trust doctrine.

A "People's Judicial Panel" listened to the testimony and provided commentary on its significance. The judges concluded that

> the governments of the world have a duty to protect the atmosphere that belongs in common to the world's people. Based on the evidence we have heard here today, the nations of our world are in violation of their most fundamental legal and constitutional obligations. They are violating the most fundamental rights of their own people and the people of the world. Each government should be legally compelled to halt its contribution to climate destruction.

The judicial panel observed that failure of governments to protect human rights and the public trust has prompted the co-owners of the atmospheric commons to turn to mass civil disobedience to protect their common property. These actions must be seen "not as violations of the law, but as attempts to enforce it." They represent the effort of tens of thousands of people to assert their collective right and responsibility to protect our public trust property, which includes resources like the earth's climate that are essential to our survival.

> Based on the evidence we have heard today, those who blockade coal-fired power plants or block tar sands oil pipelines are committing no crime. Rather, they are

exercising their right and responsibility to protect the atmospheric commons they own along with all of present and future humankind. They are acting to prevent a far greater harm—indeed, a harm that by virtue of the public trust doctrine is itself a violation of law on a historic scale.

In 1967 Bertrand Russell and Jean-Paul Sartre, undoubtedly the most famous philosophers of their era, convened the International War Crimes Tribunal, in which a distinguished panel heard evidence that the United States was committing war crimes in Vietnam. The tribunal led millions of people around the world to question the Vietnam War and encouraged tens of thousands to resist it. It has inspired many civil society tribunals since, including more than twenty independent international tribunals held in countries around the world to examine the criminality of the Iraq war. The Climate Justice judicial panel recognized that such tribunals do not have the power to force governments to comply with their orders, but it cited the view of professor of international law Richard Falk that "when governments and the UN are silent," tribunals of concerned citizens "possess a law-making authority."[2]

The Climate Justice Tribunal established a credible case that the governments and corporations of the world are systematically violating human rights, international law, and their duty to protect the public trust by allowing the greenhouse gas emissions that are destroying the earth's climate. Future climate tribunals could examine the evidence in greater detail, drawing on testimony from scientists and legal experts as well as those directly affected by climate change. They could issue declaratory judgments and injunctions. They could also make findings on the rights and responsibilities of global citizens to enforce the law and their legal rights vis-à-vis governments that try to subdue them when they do so.

Such tribunals might influence official courts to enforce the law against fossil fuel corporations and governments that are acting like their captives. And if they do not, peoples tribunals may have a role to play in legitimating the world's people to step in and engage in nonviolent constitutional insurgency to save the planet Earth.[3]

Developing the Power to Protect the Public Trust

Independently or in tandem with such tribunals, citizens can monitor violations of public trust rights and halt them through direct action. We are familiar with participatory environmental monitoring efforts like the Audubon Society's Christmas Bird Count that mobilize thousands of volunteers to collect environmental data.[4] We are also familiar with transnational teams monitoring elections around the world. Citizen plane spotters exposed the secret CIA airplanes that carried captives to rendition and torture centers. Less familiar are a number of "citizen weapon inspection teams" that have attempted to investigate nuclear weapons sites from Bangor, Maine, to Kleine Brogel in Belgium—with the teams sometimes facing arrest when they tried to cross national borders.[5] Similar local and transnational teams can monitor and expose GHG pollution as a violation of the atmospheric public trust. If they are arrested for trespass while investigating, it is ipso facto civil disobedience in defense of the public trust; and since they are attempting to exercise functions that are the duty of governments, their action is ipso facto part of a constitutional insurgency. Yet their actions are simply an attempt to protect communities and humanity against destruction of the basis of life.

Forcing governments and other actors to mend their ways will take global action on a large scale. As many as 15 million people around the world joined a day of protest against the US attack on Iraq, but the attack went forward anyway. Ultimately, a climate insurgency will require ongoing commitment from perhaps ten or twenty times that many participants worldwide. For that reason, its actions will need to be conducted in a way that can win and maintain wide public support over the long haul. A climate insurgency may need to fill the jails and make societies ungovernable through sustained disruption, but it will need to do so through nonviolent action that is increasingly perceived by the public as embodying its own deepest needs and interests.

Such a nonviolent insurgency will need to isolate and overcome specific climate-threatening institutions—coal-fired power plants, fossil fuel corporations, banks and investment houses, and government agencies that authorize and protect them. It will need to undermine the power of these institutions by utilizing their dependence on their "pillars of support"— ranging from universities and municipalities that invest their money in fossil fuel corporations to the legal authorities that order the arrest of protestors blocking pipeline construction. It will need to draw together and coordinate multiple constituencies across boundaries of nations, cultures, and beliefs. This movement will need to start by forcing incremental changes while expanding the process, consolidating and broadening the changes, and rallying wider support. Ultimately, this nonviolent insurgency will need to change the dynamics both of individual nations and of the world order.

Noninsurgent Allies

Many individuals and institutions may oppose climate change but decline to join a climate insurgency, especially at first. Some may reject its means or ends; others may agree with them but choose instead to work "within the system." Both can nonetheless be crucial allies for a climate insurgency and play a critical role in protecting both the atmosphere and the movement. A climate insurgency needs to make synergistic coordination with noninsurgent allies a strategic objective.

"Secondary institutions" such as schools, universities, religious congregations, unions, parent groups, municipal governments, neighborhoods, and workplaces can be crucial venues for such alliances. The campaign for divestment from fossil fuel companies is an example of how people can organize for climate protection within their institutions without having to take an insurgent stand themselves. Such action nonetheless undermines the legitimacy of climate destruction and those who support and permit it, and thereby lays the groundwork for the sudden crystallization of a radically new consensus for climate protection.

Those who work directly to make their own communities and institutions climate-safe through installing solar panels, insulating houses, recycling waste, and performing myriad other activities are also important allies. These people show at a grassroots level that climate-protecting change, far from being something threatening, can contribute here and now to a better life. They provide experiments in what will make GHG reduction really work for ordinary people.[6]

An insurgent movement can also find de facto allies within national political arenas and even within governments. As people react to the enormity of climate change—and to the moral challenge emanating from insurgents' acts of conscience—some individuals in all walks of life and social positions will grasp the necessity of countering it. The insurgency needs to recognize the importance of their efforts. That does not mean the movement needs to compromise with halfway measures that do not solve the problem. Indeed, it can best support those working "inside the system" by holding up the standard of what is truly necessary, even while encouraging those who are taking lesser measures.[7]

World Order Dynamics

A core strategic objective for climate protection should be to foment a competition among countries and corporations to radically reduce their GHG emissions. The movement against nuclear weapons and nuclear testing provides a significant parallel. The "peace race" is described at length in Lawrence Wittner's magisterial three-volume history, *The Struggle Against the Bomb.*[8] According to Wittner, "Most government officials—and particularly those of the major powers—had no intention of adopting nuclear arms control and disarmament policies. Instead, they grudgingly accepted such policies thanks to the emergence of popular pressure." Confronted by "a vast wave of popular resistance" they concluded, reluctantly, that "compromise had become the price of political survival." Consequently "they began to adapt their rhetoric and policies to the movement's program."

The "ban the bomb" movement demanded more of Cold War rivals than lip service or courtship. It demanded—from both sides—unilateral initiatives for peace, an end to nuclear testing, a halt to the arms buildup, and binding disarmament agreements. As Wittner massively documents, the international movement and world public opinion induced rival nations and blocs to accept the nuclear test ban treaty, détente, arms control, and the unacceptability of using nuclear weapons.

Indeed, the antinuclear movement created a bidding war among the world's leading powers. For example, as antinuclear demonstrations swelled in America and Europe in the early 1980s, arch-cold-warrior President Ronald Reagan told his astonished secretary of state, "If things get hotter and hotter and arms control remains an issue, maybe I should go see [Soviet leader] Andropov and propose eliminating all nuclear weapons." The United States thereupon agreed to forego deployment of medium-range missiles in Western Europe if Russia would remove its medium-range missiles from Eastern Europe.

When Gorbachev came to power in the USSR, he met with world peace movement leaders and unexpectedly agreed to support the peace movement's proposal for a nuclear freeze. The United States thereupon cut back on its proposed MX missiles from two hundred to fifty; abandoned plans to deploy the neutron bomb in Western Europe; and accepted the limits of the unratified SALT II arms control treaty. Ultimately, the superpowers negotiated the Strategic Arms Reduction Treaty (START), which resulted in the removal of about 80 percent of all existing strategic weapons. While these results were not fully adequate to protect the world from nuclear holocaust, they illustrate the dynamics that an independent global movement can use to force governments to move toward the long-term common interests of humanity.

The international climate protection movement seemed to be creating momentum for a similar competition among nations to cut GHG emissions. A dramatic example was the last-minute reversal, under tremendous pressure from countries and people around the world, of US opposition to the Bali Roadmap for reaching a global agreement. At Copenhagen, however, the world's two largest carbon emitters, the United States and China, joined hands to say no to the global clamor for a binding climate agreement and to take climate negotiations out of the UN venue where global pressure could be directly applied.

With the downgrading of the UN climate negotiating process, there are now few venues where such global pressure can be applied. A leaf might be taken here from the annual and biennial International AIDS Conferences, which bring thousands of scientists, public health officials, healthcare providers, advocates, affected communities, and people with AIDS together in cities around the world. In addition to releasing new scientific information and policy proposals, the highly publicized conferences set goals, targets, and standards for national AIDS policies; provide independent evaluations; shame violators; and lend international support for activists. An international climate venue modeled on the International AIDS Conference could be one way to reestablish the bidding war among climate players.[9]

A related strategic objective for a climate insurgency is the construction of a "coalition of the willing"—an alliance of countries that are prepared to take effective climate action themselves and pressure others to do the same. Such coalitions have been effective in the past, for example, the alliance of hundreds of NGOs and many governments that in 2002 created the International Criminal Court. Such a coalition was clearly visible at Copenhagen, led by the G-77 of developing nations, NGOs, and social movements from around the world, with ambiguous support from the European Union. A global insurgency and its allies can pressure countries to participate in such a climate protection coalition. The enforcement of public trust duties might serve as a guiding principle for such a configuration. The UN General Assembly might serve as a possible venue for legitimating the public trust framework.

Ultimately, countries that continue to commit waste against the global public trust can be made the target of nonviolent sanctions, implemented by willing countries and by the insurgency itself. Lord Stern has warned that the US may "have to start worrying about being shut out of markets because their production is dirty," facing restrictions or even bans on exports by 2020 due to excessive GHG emissions.[10]

The goal of this process should be not just changes in national policies, but global agreements embodying the principle of protecting the global commons and the policies necessary to realize it. Unlike past climate negotiations, however, such agreements will embody changes that already have been fought for, won public support, and been at least partially realized at the local and national level. They will embody a de facto transformation of the world order to recognize and embody the overriding authority of common human preservation.

CHAPTER 11

OVERCOMING THE OBSTACLES
TO CLIMATE PROTECTION

The global climate protection movement has initiated many of the elements necessary to overcome the obstacles to climate safety we identified in Chapter 5, "Why Climate Protection Has Failed." Others have been suggested in the preceding chapters. How can they work together as a strategy to counter climate change?

Overcoming the World Order Obstacles

Fossil Fuel–Producing Industry

The most direct sources of climate destruction are the corporations that extract and burn fossil fuel. The climate protection movement has begun to take on the fossil fuel industry. It has campaigned to shut down coal plants; opposed fossil fuel subsidies; blockaded fracking, tar sands, and offshore drilling sites; and initiated a fossil fuel divestment campaign to stigmatize fossil fuel companies and those who invest in them.

There are several additional steps that will be crucial. First, the movement needs the public trust framework or an equivalent to provide legal grounds to challenge the right of the fossil fuel companies to lay waste to the public trust. Second, it needs concrete plans for a just transition for those who are adversely affected by the rapid contraction of the fossil fuel industry. Third, it needs a strategy for energy alternatives to counter dependence on fossil fuels. This involves short-term rapid replacement like that already going on in Germany. It also requires national climate action plans like those described in Chapter 8, "Making a Country Climate-Safe," that shut down fossil fuel production and use while directing massive resources into replacing them.

Network of Support for Fossil Fuels

Behind the fossil fuel industry lies a network of supporting institutions. There are a variety of ways in which the "pillars of support" for the fossil fuel industry can be undermined. The fossil fuel divestment campaign uses a classic tactic to force universities, governments, religious organizations, and civil society groups to face up to their own complicity in global climate destruction and begin dissociating themselves from it. Organizations like Ceres and the Global Reporting Initiative establish standards for progressive greenhouse gas reduction, which the movement and allies can demand individual corporations must meet. Companies that talk the climate protection talk but don't walk the walk must be pressured to go beyond a public relations response. Pro–fossil fuel corporations and business organizations like the US Chamber of Commerce must be pressured to break away from fossil fuel industry hegemony and act on their own interest in reversing climate catastrophe.

Some unions, deeply dependent on fossil fuels, have exercised hegemony over the US labor movement as a whole. However, both the immediate interests of most unions and the general interest of working people as a whole lie with climate protection. Ending labor's support for the fossil fuel industry will withdraw a crucial pillar of political support. A program like that laid out in Chapter 8, "Making a Country Climate-Safe"—providing a just transition to protect fossil fuel–producing and –using workers, setting up a Nordic-style safety net, massively expanding climate-protecting jobs, and moving to a full employment economy—can expand the basis for union and working-class support for climate protection.

Neoliberalism

Neoliberalism has not only promoted climate-destroying industry, it has blockaded the effective use of public authority to block greenhouse gas emissions, maintaining that the habitability of the earth should be left to the market. Neoliberal ideology can be undermined by making clear its role in creating and perpetuating climate destruction. If global warming is, as Lord Stern put it, the greatest market failure in world history, then neoliberalism's promotion of unregulated markets must take much of the blame. Neoliberalism has encouraged the proliferation of GHG-promoting industries and carbon-intensive consumption around the world on a historic scale. Most important, it forms a barrier to effective climate action in the present and future. The tenets of neoliberalism must be systematically transgressed to protect our species' future.

Neoliberalism has many other destructive consequences besides destruction of the climate: It has aggravated the world's growing inequality, deepening economic stagnation, environmental degradation, and democracy deficit. In the past few years, neoliberalism and its austerity economic policies have been the target of popular upheavals around the world. These movements, however, have had little ability to reshape the global economy because they are largely limited to individual countries and lack a strategy for transforming the world order in which neoliberalism is embedded. The alternatives to neoliberalism proposed in Chapter 8, "Making a Country Climate-Safe," and Chapter 9, "A Global Trust Fund for the Global Public Trust," are oriented first of all to climate protection, but they also can open a pathway toward a confluence of the climate movement and other opponents of neoliberalism.

Nation-State System

Under the system of sovereign nation-states, each government can permit climate-destroying greenhouse gas emissions even if they destroy the earth for everyone. The key to challenging the system of absolute nation-state sovereignty is the emergence of a global climate protection movement independent of any nation or combination of nations. Such a movement does not need to abolish nations, but it does need to impose limits on the freedom of nations to destroy the common heritage of humanity. Its independence lets it define standards that all nations must follow, and its willingness to disobey established law through insurgent direct action gives it a means to impose global human interests against the will of existing states. It can thereby initiate a competitive bidding war among nations, a race to the top in which nations must take effective measures for climate protection or face delegitimization and disruption from both their own and the world's people.

The public trust doctrine can be a crucial vehicle for challenging the right of nations to lay waste to the atmosphere. It establishes duties to a nation's own people, to future generations, and to other governments and their peoples. This doctrine therefore contradicts the basic premises of absolute nation-state sovereignty.

While the movement must remain independent of any combination of nations, it can have as a strategic objective the construction of a coalition of countries that are dedicated to climate protection, both through their own actions and actions they impose on other countries. As they reduce their own GHG emissions, they can collectively function as a shadow United Nations, imposing nonviolent sanctions on nations that continue to lay waste to the common heritage of humanity. Climate protection can ultimately open a pathway that can be extended to impose global human interests in other areas as well.

Overcoming the Obstacles to Climate Protection in Human Hearts and Minds

Denialism

It is hard to promote climate protection if people are persuaded that climate change is not real, that it is not caused by human action, or that it does not present a clear and present danger to the well-being and future of individuals and of humanity as a whole. The chain that connects the release of greenhouse gases to the warming of the atmosphere to a strange assortment of devastating and sometimes paradoxical effects—including both floods and droughts and both ice melt and blizzards—is not intuitively obvious. Nor is the driving of the whole process by the political and economic forces and structures we have delineated. Challenging the denial of climate change requires more than simply asserting the authority of climate scientists; it requires making these connections in people's minds. The 350.org "connecting the dots" campaign with educational events at diverse sites affected by climate change—and at institutions perpetuating it—represented a good example of how to do so. Connecting the dots gets easier as the science gets clearer and facts are directly experienced. This process is cumulative: Many who didn't "get it" about Hurricane Katrina got it after Superstorm Sandy.

There are several grounds for encouraging people to move beyond climate denialism: That climate change denial is a deliberate strategy of manipulation by the fossil fuel companies and their allies to prevent us from protecting ourselves against their depredations. That denying the problem prevents us from taking measures to address it. That it leaves us in perpetual fear, not only of climate change but of having our climate change–denying worldview collapse in the face of reality. Recognizing climate change and acting on that recognition will feel better in the end than living a life of denial.

Incrementalism

As Bill McKibben has often pointed out, we are used to dealing with a political system based on gradualism and negotiations to reach compromise with opponents. The laws of physics don't negotiate, however. Climate change will continue on its devastating course until GHG emissions are sharply reduced. "Compromise" between the sharp reductions in GHG emissions that are necessary and continuation of business as usual will lead only to catastrophe.

Twenty-five years ago a gradualist approach might have had some credibility, but having frittered away a quarter of a century, we are like people racing toward a cliff whose only option for survival is to slam on the brakes whatever the risks. The strategy of climate insurgency provides a way not only to point out the cliff, but to focus peoples' attention on where the brakes are.

Economic Consequences of Climate Protection

Fear of the economic impact of climate protection is a significant reason that people oppose it. The climate protection movement has tended to answer concerns about negative economic effects of climate protection with the general argument that "clean energy" produces more jobs than fossil fuels. While the evidence for this argument is strong, it often fails to allay peoples' fears.

Overcoming these fears requires action on several levels. First, climate protection advocates must also become advocates for a "just transition" for those who may be negatively affected by the policies they advocate. Instead of simply saying that there will be more jobs, the movement must support specific plans to provide for the livelihoods and well-being of those in the fossil fuel–producing and –dependent industries whom its policies may put at risk.

Second, the movement needs to incorporate concrete plans for job creation in its short-term work. Energy policies in federal, state, and municipal governments and civil society need to be jobs programs as well.

Third, plans for conversion to a climate-safe economy must include the kinds of plans for full employment, rapid training, and a Nordic-style safety net described in Chapter 8, "Making a Country Climate-Safe." To win wide and continuing public support, climate protection must create a better life for ordinary people at the same time it is protecting the planet.

Let Another Country Pay

The desire to gain the benefits of climate protection while letting others pay their costs has stymied climate protection agreements and polluted climate politics. Many of the proposals in this book are designed to transform the dynamic that tempts each country to offload the costs of climate protection onto others. The crucial starting point is a global movement independent of any nation or alliance of nations. Such a movement can uphold consistent standards and make consistent demands on every nation. By participating in it, people are pressuring every other government as well as their own.

The public trust principle can also help undermine the legitimacy of the predilection to "let another country pay." Each government has an obligation to its own people to prevent waste to the atmosphere within its sphere of control. It also has an obligation to its people to prevent such waste by other governments and private parties as well. Moreover, each government, as co-trustee, has an additional obligation to the people of every other country. The public trust principle changes the question from the cost-benefit equation for each country to what each country must do to meet the obligation it owes to both its own people and the people of the rest of the world.

The Greenhouse Development Rights Framework, when combined with the public trust approach, provides another part of the solution to the "let another country pay" problem. Using the principles that would apply in a public trust law case, it lays out the share of the remedy for waste that each country is legally responsible for. It apportions that cost according to widely accepted principles of justice, rather than through the jostle of international power politics.

Finally, the proposed regime of international macro-economic cooperation is designed to reduce "beggar your neighbor"–type competition and increase mutual benefit in the global economy. By creating adequate global demand (in particular for climate-protecting goods and services), it reduces the need for countries to compete by imposing austerity on their own people or other countries. It thereby reduces the saliency of economic rivalry and gives countries an interest in each others' economic well-being. At the same time, it allows all countries to benefit economically from the transition to a climate-safe economy.

Legitimacy of the Status Quo

People are often reluctant to challenge those who are accepted as exercising legitimate authority, even when they disagree with them. A constitutional insurgency is designed to under-mine acceptance of illegitimate authority. It focuses attention on the failure of the current authorities to obey the law and fulfill their duties. In the case of climate protection, the duties of governments to protect the public trust—and their failure to do so—is the basis for challenging the authority of those who are destroying our world.

Fear of Social Movements

People often feel threatened by social movements, fearing their impact both on their own lives and on the stability of their society. The climate protection movement has consistently followed the path of nonviolence, thereby maximizing its ability to appeal to a wider pubic while minimizing fears of disruption.

One of the advantages of a nonviolent movement is that it reduces the physical threat of disruption to the wider public and those who think of themselves as bystanders, while increasing the political and emotional pressure for people to take a stand. It minimizes the ability of the authorities to portray protesters as a violent or terrorist threat. Conversely, it means that when the authorities do turn to repressive violence, they are likely to undermine their own legitimacy with their own population—to paint their own portrait as threatening, violent disruptors. It was the nonviolence of civil rights protestors that made Birmingham, Alabama's Bull Connor into an international symbol of racist oppression.

Defining a movement as a constitutional insurgency also provides a vehicle for allaying public fears of out-of-control disruption. It defines the objectives of the movement as in accord with the basic principles of democracy and constitutionalism, even while making a forthright challenge to governments that have strayed from these principles. It also embodies a commitment that the movement itself recognizes and is bound by fundamental principles of justice, democracy, and human rights and will avoid violating them in its own actions.

Individualism

The belief in the ethic of "looking out for number one" and the possibility of "getting ahead" as an individual often form a barrier to joining with others to address shared problems. The climate protection movement can help counter fear of and opposition to collective action in several ways. First, it can make clear that no individual can protect themselves, their families, and whatever else they hold dear from the effects of climate change—self-protection is an illusion. Second, the movement needs to maintain a quality of welcoming friendliness and respect for individual freedom and difference that reduces the threat felt by people who are not natural "joiners." Third, it needs to demonstrate efficacy, even if at first only in small ways. It needs to manifest enough self-discipline to belie the idea that movements represent a kind of disorder that can never realize anything worthwhile. Finally, the movement needs to provide short-term benefits that make life better for its participants. These benefits may be as simple as a friendly social atmosphere, informal transportation or childcare, entertainment, opportunities to utilize and expand personal capacities, good food, and good fellowship. The movement needs to demonstrate that being part of the movement in itself means a better life.

Hopelessness

It is easy to despair that there is anything we can do about climate change. An effective climate movement must start from a recognition, rather than a denial, of the realities that generate such despair. At the same time it proposes fellowship in action rather than moping in isolation as the best antidote to such despair. By giving people a chance to participate in meaningful action, it facilitates an alternative attitude. The strategies laid out in *Climate Insurgency*, while by no means providing a guaranteed path to climate safety, do challenge the view that climate protection is no longer worth pursuing. There is indeed much of value that it is already too late to save. But there is too much left worth saving to make the struggle for climate protection futile or irrational.

CONCLUSION

The starting point for climate protection is the recognition that climate change poses an existential threat to our species, to every individual, and to all that any of us hold dear. Climate protection is a universal existential necessity. It is blocked, however, by the way we organize our life on earth—by our world order. Conventional strategies based on the institutions of the existing world order have failed for a quarter of a century.

In response, an independent, global climate movement has arisen. It pursues the objectives and presses the demands of human survival without subservience to the interests of any one nation or coalition of nations. It is based on grassroots self-organization around the world. When conventional approaches fail, it is prepared to use mass civil disobedience.

The public trust doctrine can provide a strong legal underpinning for this movement, as it maintains that the atmosphere is the common property of present and future generations. All governments have the highest level of duty to protect it as a public trust and prevent its being wasted either by other governments or by third parties. The public trust doctrine, combined with the Greenhouse Development Rights Framework, provides a way to clearly define the just duties of each country. The climate protection movement can validly argue that governments are in violation of this duty and that citizens have the right and responsibility to enforce the protection of the atmosphere against climate destruction. Civil disobedience to protect the planet against global warming is an act of law enforcement against governments that are complicit with wasting of the atmosphere.

The effort to enforce the law against unlawful governments by means of nonviolent civil disobedience constitutes a global constitutional insurgency. It challenges the legitimacy of all governments on the basis of their failure to meet their highest duty, to protect the public trust, and it refutes the claims of polluters that their legal property rights authorize them to go on destroying the earth's climate.

Such insurgent actions can be mutually supportive with other forms of action. They can stimulate those who choose to work within established channels like electoral politics and lobbying to recognize what is necessary for genuine climate protection, even while they fight for measures that go only a small way toward solving the problem. The insurgency can encourage those who are acting here and now in civil society to convert their own lives and communities to a climate-safe basis. These "inside" and "outside" forces can be linked by coordinating networks that make their efforts synergistic and focus their power on institutions that need to be changed.

As nations are forced to recognize and act on their public trust duties, they will require "climate action plans" for the transformations they must realize. These will require a wide array of techniques for change, including government planning, investment, and regulation; decentralized civil society initiatives; and market-based incentives. Nations will also need broader changes in employment, social safety net, and other policies. Such changes will require powerful government agencies, adapted to national circumstances, to implement and oversee them. Such government authority in turn will require effective vehicles of legal, institutional, and popular power to ensure that these agencies fulfill their mission while remaining accountable to democratic institutions and the people.

While each nation must meet its own public trust responsibilities, protection of the atmosphere is inherently global. Each nation, in order to meet its own public trust responsibilities to its own people, must also ensure that all other nations, as co-trustees, also halt waste to the atmosphere. The global climate protection movement provides a vehicle by which the people of the world can insist that not only their own nation but all nations meet their obligations.

These changes do not require the abolition of the nation-state or capitalism, however desirable or undesirable that might be. But they do require substantial change in their current form. They require significant limits on the absolute freedom of states to do whatever they wish. They require significant limits on the property rights of economic actors. Ultimately these changes will need to be embodied in binding agreements expressing public trust principles. They will require pan-human institutions that can impose common human interests and necessities on all parties. Realizing and protecting these changes will take a continuing organization and mobilization of global people power. Taken together, these changes embody the transformation of the world order that is necessary to ensure our common preservation.

NOTES

Introduction

1. "Global Carbon Emissions Set to Reach Record 36 Billion Tonnes in 2013," Tyndall Centre for Climate Change Research at the University of East Anglia, November 19, 2013. http://www.uea.ac.uk/mac /comm/media/press/2013/November/global-carbon-budget-2013, accessed September 18, 2014.

2. Kiley Kroh, "Global Temperatures in April Tied for the Hottest on Record," *climateprogress*, May 20, 2014. http://thinkprogress.org /climate/2014/05/20/3439571/april-temperatures-warmest-on-record/, accessed September 18, 2014.

3. "UN Analysis: What Copenhagen Emissions Cuts Mean for Future Temperatures," *Guardian*, December 17, 2009. http://www .theguardian.com/environment/2009/dec/17/copenhagen-emissions -cuts-future-temperatures, accessed September 18, 2014.

4. Jeremy Brecher, "The Opening Shot of the Second Ecological Revolution," *Chicago Tribune*, August 16, 1988. http://articles.chicagotribune .com/1988-08-16/news/8801230163_1_greenhouse-effect-poisons -forests. See also Jeremy Brecher, "A Second Ecological Revolution?" *Common Dreams*, August 27, 2013. https://www.commondreams.org/view /2013/08/27-1. Both accessed September 18, 2014.

5. Jeremy Brecher, *Save the Humans?: Common Preservation in Action* (Boulder, CO: Paradigm Publishers, 2012).

Chapter 1

1. Murray Bookchin, "Ecology and Revolutionary Thought," originally published in 1964 in Bookchin's newsletter *Comment* and reprinted in Murray Bookchin, *Post-Scarcity Anarchism* (Berkeley, CA: Ramparts Press, 1971), p. 22.

2. See Mary E. Pettenger, ed., *The Social Construction of Climate Change* (Hampshire, England: Ashgate, 2013).

3. Spencer Weart, "International Cooperation." http://www.aip. org/history/climate/internat.htm, accessed September 19, 2014. Extensive information and interpretation on the history of climate science and policy

is maintained by physicist and historian Spencer Weart at http://www.aip .org/history/climate/index.htm. For inside accounts of the development of climate science and policy from the beginning through the implementation of the Kyoto Protocol, see Stephen H. Schneider, *Science as a Contact Sport: Inside the Battle to Save Earth's Climate* (Washington, DC: National Geographic, 2009) and Bert Bolin, *A History of the Science and Politics of Climate Change: The Role of the Intergovernmental Panel on Climate Change* (Cambridge: Cambridge UP, 2007). The accounts in this and the next chapter draw substantially on these sources.

4. Weart, "The Public and Climate Change." http://www.aip.org /history/climate/public.htm#L000 and http://www.aip.org/history /climate/public2.htm#L000, accessed September 19, 2014.

5. Naomi Oreskes and Erik M. Conway, *Merchants of Doubt: How a Handful of Scientists Obscured the Truth on Issues from Tobacco Smoke to Global Warming* (New York: Bloomsbury Press, 2011).

6. See Weart, "The Public and Climate."

7. See Weart, "The Public and Climate."

8. For discussion of some of these interpretations, see Clive Hamilton, *Requiem for a Species: Why We Resist the Truth about Climate Change* (London: Earthscan, 2010) and Weart, "The Public and Climate," pp. 3, 17, and 55.

9. John Immerwahr, "Waiting for a Signal: Public Attitudes toward Global Warming, the Environment and Geophysical Research," a report from Public Agenda, April 15, 1999. http://research.policyarchive. org/5662.pdf. Accessed September 19, 2014.

10. Matt Daily, "Exxon CEO Calls Climate Change Engineering Problem," *Reuters*, June 27, 2012. http://www.reuters.com/article/2012 /06/27/exxon-climate-idUSL2E8HRA8K20120627, accessed September 18, 2014.

11. For a discussion of the severe consequences of growing energy scarcity and their interaction with climate change, see Craig Collins, "Meet Cannibalistic Capitalism: Globalization's Evil Twin," *TruthOut*, July 30, 2012. http://truth-out.org/opinion/item/10572-meet-catabolic -capitalism-globalizations-evil-twin. See also "Cannibalistic Capitalism and Green Resistance," *TruthOut*, August 31, 2012. http://truth-out.org /news/item/11173-cannibalistic-capitalism-and-green-resistance. Both accessed September 18, 2014.

Chapter 2

1. See Ross Gelbspan, *The Heat Is On* (New York: Perseus, 1997), for a journalistic investigation of these activities. See also Weart, "The Public and Climate," p. 46.

2. Craig Collins, "Climate Change Negotiations Polarize," *Ambio* 20 no. 7, November, 1991. http://www.jstor.org/discover/10.2307/4313856?uid=3739576&uid=2129&uid=2&uid=70&uid=4&uid=3739256&sid=21103818749833, accessed September 18, 2014.

3. Craig Collins, *Toxic Loopholes: Failures and Future Prospects for Environmental Law* (Cambridge: Cambridge University Press, 2010), pp. 181–182.

4. Weart, "International Cooperation," p. 25.

5. Collins, *Toxic Loopholes*, pp. 181–182. Collins contrasts the "effective use of structural arm twisting" by the United States for the Montreal Protocol with the the refusal (or inability) of the European Community (the predecessor of the European Union) to employ structural leverage to induce American acquiescence to an effective climate agreement.

6. Some critics questioned the belief that the economic development policies pursued by Third World nations actually represent the interests of their people in escaping poverty, rather than the interests of Third World elites. For the differential carbon footprint of rich and poor in a Third World country, see Greenpeace India, *Hiding Behind the Poor*. http://www.greenpeace.org/india/Global/india/report/2007/11/hiding-behind-the-poor.pdf, accessed September 18, 2014.

7. For an inside advocate's account of how the Kyoto Protocol came to include an emissions trading system, see Graciela Chichilnisky and Kristen A. Sheeran, *Saving Kyoto* (London: New Holland, 2009).

8. September 2, 2002, quoted in Ross Gelbspan, *Boiling Point* (New York: Basic Books, 2004), p. 127.

9. Stephen H. Schneider, p. 155.

10. Stephen H. Schneider, pp. 174–175.

11. See the Business Environmental Leadership Council portion of the Center for Climate and Energy Solutions website: http://www.c2es.org/business/belc, accessed September 18, 2014.

12. "Charlemagne," *Economist*, November 18, 2006, p. 54, quoted in Weart, "International Cooperation," p. 31.

13. Weart, "The Public and Climate," p. 65.

14. Weart, "The Public and Climate," p. 54.

15. Weart, "International Cooperation," p. 30.

16. Collins, *Toxic Loopholes*, p. 185.

17. Collins, *Toxic Loopholes*, p. 189.

18. For an account sympathetic to the efforts at climate-protection legislation see Eric Pooley, *The Climate War* (New York: Hyperion, 2010). For a widely discussed analysis of the defeat of climate legislation in the US Congress, see Theda Skocpol, "Naming the Problem," February 14, 2013, prepared for the Symposium on the Politics of America's Fight against Global Warming. http://www.scholarsstrategynetwork.org/sites/default/files/skocpol_captrade_report_january_2013_0.pdf, accessed September 18, 2014. For a valuable critique of Skocpol, see Mark Engler and Paul Engler, "Climate of Change: What Does an Inside-Outside Strategy Mean?" *Dissent*, Summer 2013. http://www.dissentmagazine.org/article/climate-of-change-what-does-an-inside-outside-strategy-mean, accessed September 18, 2014.

19. Coral Davenport, "Large Companies Prepared to Pay Price on Carbon," *New York Times*, December 5, 2013. http://www.nytimes.com/2013/12/05/business/energy-environment/large-companies-prepared-to-pay-price-on-carbon.html?pagewanted=all&_r=0, accessed September 20, 2014.

20. Weart, "International Cooperation," p. 34.

21. Bill McKibben, *Oil and Honey: The Education of an Unlikely Activist* (New York: Times Books, 2013), pp. 141–142.

22. Helene Cooper, "Leaders Will Delay Deal on Climate Change," *New York Times*, November 14, 2009. http://www.nytimes.com/2009/11/15/world/asia/15prexy.html, accessed September 18, 2014.

Chapter 3

1. Collins, *Toxic Loopholes*, pp. 223–224. See also Weart, "The Public and Climate," p. 41.

2. Collins, *Toxic Loopholes*, pp. 223–224. For the dual role of transnational advocacy networks as vehicles for information exchange and as actors, see Margaret Keck and Kathryn Sikkink, *Activists beyond Borders* (Ithaca, NY: Cornell University Press, 1998).

3. Shannon M. Gibson, *Dynamics of Radicalization: The Rise of Radical Environmentalism against Climate Change*, dissertation, 2011, p. 88.

4. Gibson, *Dynamics of Radicalization*, p. 90.

5. See Ted Nace, *Climate Hope: On the Front Lines of the Fight*

against Coal (San Francisco: Coal Swarm, 2009) and Jeremy Brecher, "Jobs Beyond Coal," Labor Network for Sustainability, 2012. http://report .labor4sustainability.org/, accessed September 18, 2014.

6. See Chapter 7 of this book.

7. McKibben, *Oil and Honey*, 154ff.

8. Gibson, *Dynamics of Radicalization*, p. 1.

9. Gibson, *Dynamics of Radicalization*, p. 137.

10. Quoted in Gibson, *Dynamics of Radicalization*, p. 128.

11. Gibson, *Dynamics of Radicalization*, p. 171.

12. Quoted in Scott C. Byrd, *The UNFCCC and Global Climate Justice Activism: Rival Networks and Organizing Cascades*, dissertation, 2012, p. 77.

13. Gibson, *Dynamics of Radicalization*, p. 147.

14. The climate justice movement was to some extent an outgrowth of the broader environmental-justice movement, which for many years had focused attention on both the unjust visiting of pollution on poor and discriminated-against communities and the presence of such bias within the environmental movement itself.

15. Gibson, *Dynamics of Radicalization*, p. 67.

16. Gibson, *Dynamics of Radicalization*, p. 85.

17. Gibson, *Dynamics of Radicalization*, p. 62.

18. Gibson, *Dynamics of Radicalization*, pp. 75–76.

19. Gibson, *Dynamics of Radicalization*, p. 80.

20. Gibson, *Dynamics of Radicalization*, p. 60.

21. Byrd, *The UNFCCC*, p. 52.

22. Byrd, *The UNFCCC*, p. 58, quoting Climate Justice Alliance flyer, 2009. See Byrd 85ff for origins and character of Climate Justice Alliance.

23. Gibson, *Dynamics of Radicalization*, p. 3.

24. Gibson, *Dynamics of Radicalization*, p. 173.

25. Gibson, *Dynamics of Radicalization*, p. 3.

26. Email quoted in Gibson, *Dynamics of Radicalization*, p. 176. The quoted activist noted, "The entire climate justice movement is a compromise, is a coming together of different political cultures across their own 'business as usual,' across their own comfort zones."

27. Gibson, *Dynamics of Radicalization*, p. 3.

28. Gibson, *Dynamics of Radicalization*, p. 213.

29. There have been no post-2011 entries on the website. http:// pwccc.wordpress.com/.

30. Byrd, *The UNFCCC*, p. 130.

31. The *Climate Justice Now!* website (http://www.climate-justice
-now.org/) appears to have been largely inactive since 2012.

32. Byrd, *The UNFCCC*, p. 55.

33. Byrd, *The UNFCCC*, p. 82.

34. Byrd, *The UNFCCC*, pp. 86–87.

35. Byrd, *The UNFCCC*, p. 80. Byrd defines "prognostic framing" as "specific remedies or solutions and the general means or tactics
for achieving these objectives," p. 69.

36. "Unfortunately, these disputes create serious discord within
and between the activist and equity coalitions. This disunity undermines
efforts to build the type of North South NGO alliance among environmentalists and climate justice advocates that played such an instrumental
role in overcoming the equity road block during ozone negotiations. The
South's diminished bargaining position cannot benefit from this situation."
Collins, *Toxic Loopholes*, p. 217.

37. *The Greenpeace Chronicles: 40 Years of Protecting the Planet* (Amster-
dam: Greenpeace International, 2011), p. 126. http://www.greenpeace
.org/international/Global/international/publications/other/Greenpeace
-Chronicles.pdf, accessed September 18, 2014.

38. Stacy Morford, "NASA's James Hansen, 28 Activists Arrested
Protesting Mountaintop Mining," *InsideClimate News*, June 23, 2009.
http://insideclimatenews.org/news/20090623/nasas-james-hansen-28
-activists-arrested-protesting-mountaintop-mining, accessed September
20, 2014.

39. Gibson, *Dynamics of Radicalization*, p. 121.

40. See http://en.wikipedia.org/wiki/Camp_for_Climate_Action
#List_of_Camps_for_Climate_Action, accessed September 18, 2014.

41. Gibson, *Dynamics of Radicalization*, p. 94.

42. Gibson, *Dynamics of Radicalization*, pp. 129–130.

43. Kristin Moe, "Timeline of the Climate Movement: How
Direct Action Took Center Stage," *YES!*, October 2, 2013. http://www
.yesmagazine.org/planet/timeline-of-climate-movement-how-direct
-action-took-center-stage, accessed September 18, 2014.

44. Bill McKibben, *Fight Global Warming Now* (New York: St.
Martin's Griffin, 2007), pp. 25–26. This book by the organizers of Step
It Up gives a vivid picture of how the event was organized.

45. McKibben, *Fight Global Warming Now*, p. 36.

46. McKibben, *Fight Global Warming Now*, p. 35.

47. McKibben, *Fight Global Warming Now*, p. 26.

48. "The *Ecotone* Interview with Bill McKibben," *Ecotone* 2 no.

2 (Spring 2007). http://www.ecotonejournal.com/index.php/articles /details/the_ecotone_interview_with_bill_mckibben, accessed September 20, 2014.

49. *Ecotone* interview.

50. McKibben, *Fight Global Warming Now*, pp. 110–111. For movement self-organization see also pp. 28, 34, 41, and 54.

51. *Ecotone* interview.

52. Brecher, *Save the Humans?* p. 6.

53. "10/10/10—Global Work Party," 350.org. http://archive.350 .org/campaigns/1010. Note parallels to Gandhi's "constructive program."

54. http://act.350.org/sign/save-the-delaware/, accessed September 18, 2014.

55. http://act.350.org/sign/kosovo/, accessed September 18, 2014.

56. http://indiabeyondcoal.org/, accessed September 18, 2014.

57. http://www.moving-planet.org/about, accessed September 18, 2014. This action built in part on Power Shift 2011.

58. http://act.350.org/sign/durban-delay/, accessed September 18, 2014.

59. http://act.350.org/sign/help_nasheed/, accessed September 18, 2014.

60. See Jeremy Brecher, "Occupy Climate Change," *The Nation*, April 2, 2012. http://www.thenation.com/article/166759/occupy -climate-change#, accessed September 18, 2014.

61. http://radiowave.350.org/, accessed September 18, 2014.

62. http://chamber.350.org/, accessed September 18, 2014.

63. http://act.350.org/sign/heartland, accessed September 18, 2014.

64. McKibben, *Fight Global Warming Now*, p. 135. On abandoning control and hierarchy, see also pp. 56 and 136.

65. http://campaigns.350.org/, accessed September 18, 2014.

66. McKibben, *Fight Global Warming Now*, p. 166.

67. McKibben, *Fight Global Warming Now*, p. 111. For more on the 350.org communications strategy, visit "Jamie Henn Takes Us Inside 350 .org," podcast, Jon Christensen blog. http://christensenlab.net/jamie -henn-takes-us-inside-350-org/, accessed September 18, 2014.

68. Ryan Lizza, "The President and the Pipeline," *New Yorker*, September 16, 2013. http://www.newyorker.com/magazine/2013/09/16 /the-president-and-the-pipeline, accessed September 20, 2014.

69. For background on early indigenous opposition to Alberta tar sands development see Lori Waller, "We Can No Longer Be Sacrificed," *Briarpatch Magazine,* June 9, 2008. http://briarpatchmagazine.com

/articles/view/we-can-no-longer-be-sacrificed, accessed September 18, 2014.

70. Bob Wilson, "Forging the Climate Movement: Environmental Activism and the Keystone XL Pipeline," Environments and Societies Series, University of California–Davis, November 20, 2013. http://environmentsandsocieties.ucdavis.edu/files/2011/11/Wilson-ES-11.20.2013.pdf, accessed September 18, 2014.

71. Bill McKibben, "Good News: Tar Sands Action Joined 350.org." http://tarsandsaction.org/2012/01/02/good-news-tar-sands-action-joining-350-org/, accessed September 18, 2014.

72. McKibben, Oil and Honey, p. 32.

73. Lizza, "The President and the Pipeline."

74. McKibben, Oil and Honey, p. 56.

75. Jeremy Brecher and the Labor Network for Sustainability, "Stormy Weather: Climate Change and a Divided Labor Movement," New Labor Forum, January/February 2013. http://nlf.sagepub.com/content/22/1/75.full, accessed September 18, 2014.

76. McKibben, Oil and Honey, p. 45.

77. McKibben, Oil and Honey, p. 80.

78. McKibben, Oil and Honey, p. 82.

79. McKibben, Oil and Honey, p. 87. The House subsequently passed the bill again.

80. McKibben, Oil and Honey, p. 128.

81. Lizza, "The President and the Pipeline."

82. McKibben, Oil and Honey, pp. 251, 253, and 255. See also Bob Wilson, "Forging the Climate Movement." http://environmentsandsocieties.ucdavis.edu/files/2011/11/Wilson-ES-11.20.2013.pdf, accessed September 30, 2014. 350.org, with its usual lack of turf protection, encouraged the Sierra Club to take the lead.

83. Katherine Bagley, "To Defeat Keystone, Environmental Movement Goes from Beltway to Grassroots," InsideClimate News, May 23, 2013. http://insideclimatenews.org/news/20130523/defeat-keystone-environmental-movement-goes-beltway-grassroots, accessed September 20, 2014.

84. Zoe Carpenter, "Tar Sands Blockade: The Monkey-Wrenchers," Rolling Stone, April 11, 2013.

85. Kristin Moe, "#FearlessSummer: How the Battle to Stop Climate Change Got Ferocious," YES!, September 10, 2013. http://www.yesmagazine.org/planet/fearless-summer-how-battle-to-stop-climate-change-got-ferocious, accessed September 20, 2014. See also Wen

Stephenson, "The New Climate Radicals," *The Nation*, July 17, 2013. http://www.thenation.com/article/175316/new-climate-radicals, and Wen Stephenson, "The Grassroots Battle against Big Oil," *The Nation*, October 28, 2013; http://www.thenation.com/article/176556/grassroots-battle-against-big-oil. Steve Early and Suzanne Gordon, "California Refinery Town Hits Chevron with One-Two Punch," *Counterpunch*, August 6, 2013. http://www.counterpunch.org/2013/08/06/california-refinery-town-hits-chevron-with-one-two-punch/, accessed September 20, 2014.

86. Crysbel Tejada and Betsy Catlin, "Indigenous Resistance Grows Strong in Keystone XL Battle," *Waging Nonviolence*, May 8, 2013. http://wagingnonviolence.org/feature/indigenous-resistance-grows-strong-in-keystone-xl-battle/, accessed September 18, 2014.

87. "Sign the Keystone XL Pledge of Resistance," Credo Action. https://act.credoaction.com/sign/kxl_pledge, accessed May 21, 2014.

88. Ken Butigan, "The Keystone XL Pledge of Resistance Takes Off," *Waging Nonviolence*, June 21, 2013. http://wagingnonviolence.org/feature/the-keystone-xl-pledge-of-resistance-takes-off/, and Elijah Zarlin, "200 People Went to the State Department to say NoKXL. Here's What Happened." #NOKXL. http://nokxl.org/200-people-went-to-the-state-department-to-say-nokxl-heres-what-happened/.

89. Bob Wilson, "Forging the Climate Movement," Rev. Lennox Yearwood, president of the Hip Hop Caucus interview, p. 27. http://environmentsandsocieties.ucdavis.edu/files/2011/11/Wilson-ES-11.20.2013.pdf.

90. Sarah Wheaton, "Pipeline Fight Lifts Environmental Movement," *New York Times*, January 24, 2014. http://www.nytimes.com/2014/01/25/us/keystone-xl-pipeline-fight-lifts-environmental-movement.html, accessed September 18, 2014.

91. Kevin Begos and Joann Loviglio, "College Fossil fuel Divestment Movement Builds," *AP*, May 22, 2013. http://news.yahoo.com/college-fossil fuel–divestment-movement-builds-173849305.html, accessed September 18, 2014.

92. Ellen Dorsey and Richard N. Mott, "Philanthropy Rises to the Fossil Divest-Invest Challenge," *Huffington Post*, January 30, 2014. http://www.huffingtonpost.com/ellen-dorsey/philanthropy-rises-to-the_b_4690774.html, accessed September 18, 2014. Dorsey was an activist in the campaign for divestment from South Africa.

93. McKibben, *Oil and Honey*, p. 96.

94. McKibben, *Oil and Honey*, pp. 145–149.

95. McKibben, *Oil and Honey*, p. 230.

96. Katherine Bagley, "Spreading Like Wildfire, Fossil Fuel Divestment Campaign Striking a Moral Chord," *InsideClimate News*, December 6, 2012. http://insideclimatenews.org/news/20121206/climate-change -activists-350-bill-mckibben-divestment-fossil fuels-universities -harvard-coal-oil-gas-carbon, accessed September 20, 2014.

97. McKibben, *Oil and Honey*, p. 236.

98. Michael Wines, "Stanford to Purge $18 Billion Endowment of Coal Stock," *New York Times*, May 6, 2014. http://www.nytimes .com/2014/05/07/education/stanford-to-purge-18-billion-endowment -of-coal-stock.html?_r=0, accessed September 18, 2014.

99. *Fossil Free*, "Divestment Commitments." http://gofossilfree.org /commitments/, accessed September 18, 2014.

100. Quoted in "Campaign against Fossil Fuels Growing, Says Study," *Guardian*, October 8, 2013. http://www.theguardian.com /environment/2013/oct/08/campaign-against-fossil-fuel-growing, accessed September 20, 2014.

101. Dorsey and Mott. See Brendan Smith, Jeremy Brecher, and Kristen Sheeran, "Where Should the Divestors Invest?" *Common Dreams*, May 17, 2014. https://www.commondreams.org/view/2014/05/17, and Brendan Smith and Jeremy Brecher, "Do the Math: Invest While We Divest," *Common Dreams*, November 14, 2012. https://www .commondreams.org/view/2012/11/14. Both accessed September 18, 2014.

102. John Schwartz, "Rockefellers, Heirs to an Oil Fortune, Will Divest Charity of Fossil Fuels," *New York Times*, September 21, 2014. http://www.nytimes.com/2014/09/22/us /heirs-to-an-oil-fortune-join-the-divestment-drive.html?action=clic &contentCollection=N.Y.%20%2F%20Region&module=Related Coverage&region=Marginalia&pgtype=article.

Chapter 4

1. "People's Climate March—Wrap-up," 350.org. http:// peoplesclimate.org/wrap-up/ and "Climate change summit: Global rallies demand action," *BBC News*, September 21, 2014. http://www.bbc.com /news/science-environment-29301969.

Chapter 5

1. Elizabeth Kolbert, "The Catastrophist," *New Yorker*, June 29, 2009. http://thingsbreak.files.wordpress.com/2009/06/hansennewyorker .pdf, accessed September 18, 2014.

2. Richard Heede, *Carbon Majors* (Snowmass, CO: Climate Mitigation Services, 2014). http://www.climateaccountability.org/pdf /MRR%209.1%20Apr14R.pdf, accessed September 18, 2014.

3. "Fortune Global 500," *CNN Money*. http://money.cnn.com /magazines/fortune/global500/2013/full_list/, accessed September 18, 2014. Totals calculated in Sean Sweeney, "Unions, Climate Change and the Great Inaction," unpublished article.

4. Gene Sharp, *Waging Nonviolent Struggle* (Boston: Porter Sargent, 2005), p. 35.

5. Richard Falk, "A Radical World Order Challenge: Addressing Global Climate Change and the Threat of Nuclear Weapons," *Globalizations*, March–June, 2010, p. 147. http://www.tandfonline.com/doi/abs /10.1080/14747731003593414?journalCode=rglo20#preview, accessed September 20, 2014. See this paper for a brilliant analysis of world order constraints on climate protection. "Presentism" refers to the short time horizon that guides the principal actors of the world order and a chaotic world order's structural inability to manage long-term consequences.

A different structural approach maintains that capitalism per se is the cause of climate change. See, for example, Richard Smith, "Capitalism and the Destruction of Life on Earth," *Real-World Economics Review* 64. http://www.paecon.net/PAEReview/issue64/Smith64.pdf, accessed September 18, 2014.

Such a presumption is problematic for several reasons. Whether the abolition of capitalism would be a good or a bad thing, it should not be seen as a necessary or sufficient condition for climate protection. Rapid and effective climate-protection measures can be taken in capitalist societies—for example, the transformation of much of Germany's electrical power system to renewable sources. Historically, capitalism has been able to live within limits—for example, the abolition of slavery, rule by democratic governments, trade unionism, and the welfare state. Neoliberal ideology notwithstanding, restriction of external costs to prevent market failures is not incompatible with capitalism in theory or in practice. The competition of states in a system based on nation-state sovereignty is a crucial cause of climate change that cannot be reduced to capitalism per se. Nor should it be assumed that replacing capitalist states with socialist ones will automatically fix climate change; socialist states have engaged in severe environmental destruction, and sovereign socialist states would still have incentives to continue GHG emissions. Even the most optimistic timeframe for the abolition of capitalism will leave a world devastated by climate change. While putting limits on capitalist dynamics is clearly essential for climate protection, maintaining that capitalism must be abolished before effective climate protection can be implemented is not only false but also discourages effective action to protect the climate before it is destroyed entirely.

For a more nuanced view of the role of capitalism in the climate crisis see Naomi Klein, "Capitalism vs. the Climate," *The Nation*, November 28, 2011. http://www.thenation.com/article/164497/capitalism-vs -climate?page=0,5#sthash.H9ztcu1v.dpuf, accessed September 18, 2014. Klein describes "the blindingly obvious roots of the climate crisis" as "globalization, deregulation and contemporary capitalism's quest for perpetual growth" without regarding the abolition of capitalism as the necessary or sufficient condition for climate protection.

6.   Naomi Klein, "Naomi Klein: Big Green Is in Denial," *Salon*, September 5, 2013. http://www.salon.com/2013/09/05/naomi_klein _big_green_groups_are_crippling_the_environmental_movement _partner/, accessed September 18, 2014.

7.   Economists refer to this as the "free-rider problem." However, there is more to it than wanting benefits without paying for them. The capitalist and state systems are to some extent competitive, zero-sum games, in which everyone ending up better off is not perceived as a win. Players may aim not just to avoid costs themselves, but to force costs on their rivals in order to restrict their development.

Chapter 6

1.   The closely related term "civil resistance" is sometimes used to describe a somewhat broader set of movements, of which constitutional insurgencies are a subset. *Civil Resistance and Power Politics*, by Adam Roberts and Timothy Garton Ash (Oxford: Oxford University Press, 2009), portrays a variety of such movements and observes that they are "civil" in the sense that "they had a civic quality, relating to the interests and hopes of society as a whole"; in some cases "the action involved was not primarily disobedient, but instead involved supporting the norms of a society against usurpers"; and the "generally principled avoidance of

the use of violence was not doctrinaire" (p. 4). The distinctive hallmark of insurgencies is that they reject the authority of the state. For a discussion of the distinction between civil disobedience and civil resistance see Ellen Barfield, "Defending Resistance," *WIN*, Spring 2011. http://www .warresisters.org/content/defending-resistance, accessed September 18, 2014. Barfield says, "Resistance is understood to include legally challenging the government's behavior, and urging juries and judges to uphold the citizen's right and responsibility to protest government wrongdoing by acquitting accused resisters."

While Gandhi is often portrayed as the apostle of "civil disobedience," he wrote, "I found that even 'Civil Disobedience' failed to convey the full meaning of the struggle. I therefore adopted the phrase 'Civil Resistance'" Letter to P. K. Rao, September 19, 1935, as quoted in Louis Fischer, *The Life of Mahatma Gandhi* (New York, 1962), pp. 93–94.

2. Mohandas K. Gandhi, *Indian Opinion*, November 11, 1905.

3. Jeremy Brecher, "Civil Disobedience as Law Enforcement," *Waging Nonviolence*, August 14, 2013. http://wagingnonviolence.org /feature/civil-disobedience-as-law-enforcement/, accessed September 18, 2014.

4. For the role of law in relation to social movements, see Brendan Smith, "Why War Crimes Matter," in Brecher, Cutler, and Smith, *In the Name of Democracy* (New York: Metropolitan Books, 2005), pp. 308–321.

5. For authority as based on a promise to obey, and the possibility of "extended authority" that illegitimately uses such authority for purposes for which it was not intended, see Charles Lindblom, *Politics and Markets* (New York: Basic Books, 1977).

6. See the discussions of the application of human rights law to climate and other environmental issues in Burns H. Weston and David Bollier, *Green Governance: Ecological Survival, Human Rights, and the Law of the Commons* (New York: Cambridge University Press, 2013). See also Douglas A. Kysar, "Climate Change and the International Court of Justice," Yale University Law School, August 14, 2013. http://papers .ssrn.com/sol3/papers.cfm?abstract_id=2309943, accessed September 18, 2014.

7. Jonathan Schell, "Introduction," in Adam Michnik, *Letters from Prison* (Berkeley: University of California Press, 1987).

8. James Gray Pope, "Labor's Constitution of Freedom," *The Yale Law Journal* 106, 1997. http://papers.ssrn.com/sol3/papers.cfm?abstract _id=1622865, accessed September 18, 2014.

Chapter 7

1. See Mary Christina Wood, "Atmospheric Trust Litigation across the World," in William C.G. Burns and Hari M. Godowsky, eds., *Adjudicating Climate Change: Sub-National, National, and Supra-National Approaches* (New York: Cambridge University Press, 2009). http://law.uoregon.edu/wp-content/uploads/2011/11/ATL-Across-the-World.pdf, accessed September 18, 2014. p. 104 and references there.

2. Weston and Bollier, *Green Governance*, p. 239.

3. Charles Wilkinson, "The Headwaters of the Public Trust," *Environmental Law* 19 (1989), p. 425.

4. Weston and Bollier, *Green Governance*, p. 135.

5. United Nations Environment Program, Division of Environmental Law and Conventions, "IEG of the Global Commons: Background." http://www.unep.org/delc/GlobalCommons/tabid/54404/Default.aspx, accessed September 18, 2014.

6. Lindsay Kucera, "Does the U.S. Have a Legal Responsibility to Stop Climate Change?" *YES!*, February 22, 2012. http://www.yesmagazine.org/planet/why-im-suing-the-federal-government, accessed September 18, 2014.

7. The Atmospheric Trust Litigation Project is coordinated by Our Children's Trust: http://www.ourchildrenstrust.org, accessed September 18, 2014.

8. Wood, "Atmospheric Trust Litigation," p. 153.

9. Amended complaint: http://www.eenews.net/assets/2012/06/29/document_pm_02.pdf, accessed September 18, 2014.

10. Weston and Bollier, *Green Governance*, p. 242.

11. Jeremy Hsieh, "Alaska's High Court First Supreme Court in the Nation to Hear Climate Change Case," *Alaska Public Media*, October 4, 2013. http://www.alaskapublic.org/2013/10/04/alaskas-high-court-first-supreme-court-in-the-nation-to-hear-climate-change-case/, accessed September 18, 2014.

12. Wood, "Atmospheric Trust Litigation." For more on the concept of the atmospheric public trust, see Wood's *Nature's Trust: Environmental Law for a New Ecological Age* (Cambridge: Cambridge University Press, 2013).

13. Wood, "Atmospheric Trust Litigation," p. 106.

14. Wood, "Atmospheric Trust Litigation," p. 110.

15. http://blogs.law.widener.edu/envirolawcenter/2013/12/21

/the-pennsylvania-supreme-courts-robinson-township-decision-a-step
-back-for-marcellus-shale-a-step-forward-for-article-i-section-27/.

16. Wood, "Atmospheric Trust Litigation," pp. 124–125.

17. Wood, "Atmospheric Trust Litigation," p. 126.

18. Wood, "Atmospheric Trust Litigation," p. 132.

19. James Hansen et al., "The Case for Young People and Nature: The Path to a Healthy, Natural, Prosperous Future," p. 12. http://www.columbia.edu/~jeh1/mailings/2011/20110505_CaseForYoungPeople.pdf, accessed September 18, 2014.

20. Wood, "Atmospheric Trust Litigation," 135ff.

21. Paul Baer, Tom Athanasiou, Sivan Kartham, and Eric Kemp-Benedict, *The Greenhouse Development Rights Framework* (Berlin: Heinrich Böll Foundation, Christian Aid, EcoEquity and the Stockholm Environment Institute, 2008). http://www.ecoequity.org/docs/TheGDRsFramework.pdf, accessed September 18, 2014.

22. Wood, "Atmospheric Trust Litigation," pp. 142–145.

23. Wood, "Atmospheric Trust Litigation," pp. 147–148.

24. Lawrence Hurley, "The Mother behind Kids' Long-Shot Legal Crusade," *Greenwire,* December 19, 2012. http://www.eenews.net/stories/1059974030, accessed September 18, 2014.

25. Michael Wolkind, "How We Won Acquittal of Kingsnorth Six," *Guardian,* May 31, 2009. http://www.theguardian.com/environment/cif-green/2009/may/31/kingsnorth-defence-lawyer/print, accessed September 18, 2014.

26. John Vidal, "Not Guilty: The Greenpeace Activists Who Used Climate Change as a Legal Defence," *Guardian*, September 10, 2008. http://www.theguardian.com/environment/2008/sep/11/activists.kingsnorthclimatecamp, accessed September 18, 2014.

American law includes a parallel to "lawful excuse," known as the "necessity defense." Defendants can argue that even if they violated a law, their action was necessary to prevent far greater damage. American courts have several times acquitted war protesters who made a necessity defense, but as yet they have not acted similarly for climate protesters. See Ellen Barfield, "Defending Resistance." http://www.warresisters.org/content/defending-resistance, accessed September 18, 2014.

27. Text of speech delivered by Alec Johnson in Nacogdoches, TX, on September 21, 2014, posted at https://www.facebook.com/PostCarbon/posts/10152530065808369.

28. "Stand with Climate Hawk, Alec Johnson" at https://climatehawk.wufoo.com/forms/stand-with-climate-hawk-alec-johnson and

Alec Johnson, "Blockadia on Trial: What the Jury Did Not Hear," Huffington Post, November 11, 2014. http://www.huffingtonpost.com/alec -johnson/keystone-xl-blockade_b_6134732.html.

29. Wood, "Atmospheric Trust Litigation," pp. 118–119.

30. Wood, "Atmospheric Trust Litigation," p. 142.

31. Gandhi's celebrated salt march can be seen as a constitutional insurgency against abuse of a public trust—access to the resources of the sea. The British colonial government forbade Indians from harvesting salt from the sea, established a salt monopoly, and imposed a heavy tax on its production and sale. In 1930, Gandhi led a march to the sea to harvest salt. Millions of Indians engaged in civil disobedience by harvesting salt in violation of British law; 80,000 of them were jailed. The campaign marked a turning point in the campaign for Indian independence.

Chapter 8

1. Wen Stephenson, "The New Climate Radicals," *The Nation*, August 5–12, 2013. http://www.thenation.com/article/175316/new -climate-radicals?page=full#axzz2cimwsrRq, accessed September 18, 2014.

2. James Hansen et al., "The Case for Young People and Nature: The Path to a Healthy, Natural, Prosperous Future," p. 12.

3. For a review of such studies see Laurence L. Delina and Mark Diesendorf, "Is Wartime Mobilisation a Suitable Policy Model for Rapid National Climate Mitigation?" *Energy Policy*, July 2013, section 2. http://www.sciencedirect.com/science/article/pii/S0301421513002103, accessed September 18, 2014. See also Robert Pollin, Heidi Garrett-Peltier, and James Heintz, *Green Growth: A Program for Controlling Climate Change and Expanding U.S. Job Opportunities* (Washington, DC: Center for American Progress, 2014).

4. Laurence L. Delina and Mark Diesendorf, "Is Wartime Mobilisation a Suitable Policy Model for Rapid National Climate Mitigation?" *Energy Policy*, July 2013, section 2. http://www.sciencedirect.com /science/article/pii/S0301421513002103, accessed September 18, 2014, and Laurence L. Delina and Mark Diesendorf, "Governing Rapid Climate Mitigation," January, 2013. http://tokyo2013.earthsystemgovernance .org/wp-content/uploads/2013/01/0134-DELINA_DIESENDORF.pdf, accessed September 18, 2014.

5. Unless otherwise noted, this discussion is based on Delina and

Diesendorf's "Is Wartime Mobilisation a Suitable Policy Model for Rapid National Climate Mitigation?" section 3.

6. In many countries indirect taxes were levied on consumer goods that competed with military production for resources.

7. This discussion is based on Delina and Diesendorf, "Is Wartime Mobilisation a Suitable Policy Model for Rapid National Climate Mitigation?" section 3, and additional sources as noted.

8. Private investment in fossil fuel–reducing activities has not been forthcoming even in many cases where such investments would have paid for themselves or even made a profit. A 2007 study by the McKinsey consulting firm found that the United States could rapidly cut 28 percent of its greenhouse gases at fairly modest cost and with only small technological innovations. According to the director of the study, Jack Stephenson, "These types of savings have been around for 20 years." But according to another research team member, "There is a lot of inertia, and a lot of barriers." To give just one example, if tenants pay for their heat, landlords have no incentive to buy any but the cheapest, least energy-efficient furnaces. See Matthew L. Wald, "Study Details How U.S. Could Cut 28% of Greenhouse Gases," *New York Times*, November 30, 2007. http://www.nytimes.com/2007/11/30/business/30green.html?_r=3&oref=slogin&, accessed September 18, 2014.

These findings raise doubts that policies that rely on charges for carbon emissions will in fact promote massive investment in climate-protecting activities.

9. Board of Governors of the Federal Reserve System, "Industrial Production and Capacity Utilization—G.17," September 15, 2014. http://www.federalreserve.gov/RELEASES/G17/current/default.htm, accessed September 18, 2014.

10. See Mary Christina Wood, "Recouping Natural Resource Damages," in *Nature's Trust* (New York: Cambridge University Press, 2014), p. 185ff. For a proposal for recouping natural resource damages at a global level, see Julie-Anne Richards and Keely Boom, *Carbon Majors Funding Loss and Damage* (Berlin: Heinrich Böll Foundation, 2014). http://www.boell.de/sites/default/files/carbon_majors_funding_loss_and_damage_kommentierbar.pdf, accessed September 18, 2014.

11. Bureau of Labor Statistics news release, "The Employment Situation—August 2014." http://www.bls.gov/news.release/pdf/empsit.pdf, accessed September 18, 2014. The labor force participation rate in the United States is below 64 percent, indicating a vast further labor reserve that could be tapped given conditions of continuing full employment.

12. Massive job creation for climate protection need not wait for such retraining, however. An emergency climate-protection program will require a wide range of activities drawing on diverse skills, from planting trees to developing new energy-conserving software to installing solar panels. Climate protection can learn a lesson from the New Deal's Works Progress Administration (WPA), which during the 1930s was able to rapidly employ millions and substantially reduce unemployment because of its emphasis on putting people to work doing things that utilized their existing skills.

13. For the "Nordic model," see the section by that name in Labor Network for Sustainability, "Labor, Sustainability, and Justice," August 17, 2011. http://www.labor4sustainability.org/wp-content/uploads/2011/09/labor-sustainability-and-justice.pdf, accessed September 18, 2014.

14. Mark Drajem, "Sierra Club Says 142 Coal-Fired Plants Shut During Drive," *Bloomberg*, March 1, 2013. http://www.bloomberg.com/news/2013-03-01/sierra-club-says-142-u-s-coal-fired-plants-during-drive.html, accessed September 18, 2014.

15. Osha Gray Davidson, "Germany Has Built Clean Energy Economy That U.S. Rejected 30 Years Ago," *InsideClimate News*, November 13, 2012. http://insideclimatenews.org/news/20121113/germany-energiewende-clean-energy-economy-renewables-solar-wind-biomass-nuclear-renewable-energy-transformation?page=2, accessed September 18, 2014.

16. For a fuller discussion of the arguments around economic growth, see "Labor, Sustainability, and Justice," Labor Network for Sustainability, August 17, 2011, especially the "Growth for What? Growth for Whom?" section. http://www.labor4sustainability.org/wp-content/uploads/2011/09/labor-sustainability-and-justice.pdf, accessed September 18, 2014.

Chapter 9

1. Katie Allen, "ILO Report Warns Unemployment 'a Major Global Challenge' for Years," *Guardian*, June 3, 2013. http://www.guardian.co.uk/business/2013/jun/03/ilo-report-unemployment-global-challenge, accessed September 18, 2014.

2. Richard Falk, "Opening Speech at the World Tribunal on

Iraq," June 24, 2005. http://www.wagingpeace.org/opening-speech-at -the-world-tribunal-on-iraq.

3. Livestream of the Peoples Climate Justice Tribunal is avail-able at http://new.livestream.com/TheNewSchool/peoples-climate -justice-summit. The members of the People's Judicial Panel were Lisa Garcia, Earth Justice; Julia Olson, Our Children's Trust; Rex Varona, Global Coalition on Migration; and Jeremy Brecher, Labor Network for Sustainability.

4. See Brendan Smith, Tim Costello, and Jeremy Brecher, "Green Paper Gold," *Foreign Policy in Focus*, December 9, 2008. http://fpif.org /green_paper_gold, and Jeremy Brecher, "How to Pay for a Global Climate Deal," *InsideClimate News*, March 20, 2009. http:// insideclimatenews.org/blog/523. For an intriguing precursor of this idea, see Jeremy Brecher, Tim Costello, and Brendan Smith, "Global Labor's Forgotten Plan to Fight the Great Depression," *History News Network*, March 22, 2009. http://hnn.us/article/69169. All accessed September 18, 2014.

5. "IMF Head Worried about Lack of Fiscal Stimulus," *Reuters*, December 22, 2008. http://uk.reuters.com/article/2008/12/22/business -us-financial-imf-stimulus-idUKTRE4BL3WN20081222, accessed September 18, 2014.

6. Jeremy Brecher, "How to Pay for a Climate Deal."

7. Former IMF chief economist Simon Johnson explained the proposal thus: "The principle behind it is that everyone would get bonus dollars. The objective is to create a windfall of cash." Edmund Conway, "IMF Poised to Print Billions of Dollars in 'Global Quantitative Eas-ing,'" *Telegraph*, March 13, 2009. http://www.telegraph.co.uk/finance /recession/4986287/IMF-poised-to-print-billions-of-dollars-in-global -quantitative-easing.html, accessed September 18, 2014.

As in a stimulus measure applied to a national economy (for example, the American Recovery and Reinvestment Act of 2009), if the windfall of cash is used to generate economic activity, then the value cre-ated by the new activity is what ultimately pays for the initial spending. The global recession has put millions of people and productive resources out of service. If they could be mobilized effectively, these vast unused productive capacities could help rebuild the global economy on a low-carbon basis.

Neoliberal opinion warned that paper gold would cause infla-tion, but in the midst of a historic crisis of deflation, the IMF, the United States, and the great majority of economists called for economic stimulus to counter deflation. Further, if SDRs are used to stimulate work and production through green public works using material and human resources that would otherwise lie idle, they will create new value at least as great as their own value, forestalling any inflationary effect. Even if there were an inflationary effect, it would affect all countries approximately equally, so that one of the main downsides of inflation—exchange-rate volatility—would not be increased.

It is easy to agree in principle that all countries should coordinate their economies to provide their fair share of the needed global economic stimulus, but in practice they often pursue their own national interests or those of their most politically powerful constituencies. That's why national stimulus spending carries a risk. The stimulus will create new spending at home, but in a globalized economy it may primarily benefit the economies of other nations that supply cheap exports and do not stimulate their own economies. As Tom Vosa, head of economic research at NAB Capital in London, explained, "If one or two countries do fiscal packages, that's simply going to boost the export market for countries which haven't." Paper gold, however, overcomes this because it stimulates the global economy as a whole, and therefore benefits the global economy as a whole. See John W. Schoen, "Fault Lines Open in Talks over Global Crisis Fixes," *NBC News*, March 13, 2009. http://www.nbcnews.com/id/29666007/wid/22224893/page/2/#.Ukw4nOD3DF0, accessed September 18, 2014.

8. "IMF proposes 'Green Fund' for Climate Change Financing," IMF Survey online, January 30, 2010. http://www.imf.org/external/pubs/ft/survey/so/2010/NEW013010A.htm, accessed September 18, 2014.

See also Soren Ambrose and Bhumika Muchhala, *Fruits of the Crisis: Leveraging the Financial and Economic Crisis of 2008–2009 to Secure New Resources for Development and Reform the Global Reserve System* (Penang: Third World Network, 2010).

Chapter 10

1. Jeremy Brecher, "Why Getting Arrested to Resist the Keystone XL Is Legally Justified," *Waging Nonviolence*, April 24, 2014. http://wagingnonviolence.org/feature/getting-arrested-resist-keystone-xl-legally-justified/, accessed September 18, 2014.

2. Richard Falk, "The Accountability of Leaders: A Challenge to Governments and Civil Society," in Jeremy Brecher, Jill Cutler, and

Brendan Smith, eds., *In the Name of Democracy* (New York: Metropolitan Books, 2005), p. 307.

3. Richard Falk, "Opening Speech at the World Tribunal on Iraq," June 24, 2005," http://www.wagingpeace.org/opening-speech-at-the-world-tribunal-on-iraq/, accessed September 18, 2014.

4. See Weston and Bollier, *Green Governance*, pp. 163–164, for additional examples.

5. See "Citizen Weapons Inspections," *Free Library*. http://www.thefreelibrary.com/CITIZENS+WEAPONS+INSPECTIONS.-a053683039, accessed September 18, 2014.

6. Such actions are the equivalent of what Gandhi called his "constructive program" for grassroots social development that was the complement to his campaigns of mass civil disobedience.

7. As Adam Michnik pointed out during the rise of Solidarity in Poland in his *Letters from Prison*, the best way for an independent movement to support reformers in government is to put pressure on the regime—thereby strengthening the hand of those who would make greater concessions to the insurgents. The fear that they may lose their legitimacy to insurgent challengers often provides authorities their greatest motivation for reform.

8. Lawrence S. Wittner, *The Struggle against the Bomb* (Stanford, CA: Stanford University Press, 1993–2003).

9. See Jeremy Brecher and Kevin Fisher, "Climate Protection Can Learn from the AIDS Movement," *Nature Climate Change*, September 25, 2013. http://www.nature.com/nclimate/journal/v3/n10/full/nclimate1986.html, accessed September 18, 2014.

10. Ben Webster, "Climate Change Action Countries Will Ban 'Dirty' US Exports, Lord Stern Warns," *Australian*, November 19, 2010. http://www.theaustralian.com.au/national-affairs/policy/climate-change-action-countries-will-ban-dirty-us-exports-lord-stern-warns/story-e6frg6xf-1225956283593, accessed September 18, 2014.

For background on nonviolent sanctions, see the work of Gene Sharp. For the application of nonviolent sanctions in a different context, see the section "The world says no—and yes," in Jeremy Brecher, "Terminating the Bush Juggernaut." http://www.jeremybrecher.org/peace/terminating-the-bush-juggernaut/, accessed September 18, 2014.

Index

ABOUT THE AUTHOR

Jeremy Brecher (www.jeremybrecher.org) is the author of more than a dozen books on labor and social movements, including *Save the Humans? Common Preservation in Action* and his classic labor history *Strike!*, which was just published in a revised fortieth-anniversary edition. He has been writing about climate protection strategy since 1988, is a founder of the Labor Network for Sustainability (www.labor4sustainability.org), and was arrested in the first White House protests against the Keystone XL pipeline. Over the course of half a century he has participated in movements for nuclear disarmament, civil rights, peace in Vietnam, international labor rights, global economic justice, accountability for war crimes, and many others. He holds a PhD from Union Graduate School and has received five regional Emmy Awards for his documentary film work.